THE INSIDER'S GUIDE TO

CHOOSING
AND BUYING A
YACHT

THE INSIDER'S GUIDE TO

CHOOSING
AND BUYING A
YACHT

Duncan Kent

WILEY

John Wiley & Sons, Ltd

This edition first published 2011

Copyright © 2011 John Wiley & Sons Ltd

Many photos in this book were reproduced courtesy of professional marine photographer Rick Buettner. The author and the publisher would like to credit and thank him for his contributions.

Registered office

John Wiley & Sons Ltd, The Atrium, Southern Gate, Chichester, West Sussex, PO19 8SQ, United Kingdom

For details of our global editorial offices, for customer services and for information about how to apply for permission to reuse the copyright material in this book please see our website at www.wiley.com.

Library of Congress Cataloging-in-Publication Data

Kent, Duncan.
    The insider's guide to choosing and buying a yacht / Duncan Kent.
      p. cm.
    Includes bibliographical references and index.
    ISBN 978-0-470-97269-4 (pbk.)
      1. Yachts—Purchasing—Handbooks, manuals, etc. I. Title.
    VM333.K46 2011
    623.822'3—dc22

2010046085

A catalogue record for this book is available from the British Library.

Set in Size of 10/12 and Humanist 777 BT by MPS Limited, a Macmillan Company, Chennai, India

Printed in Great Britain by Bell and Bain, Glasgow

At Wiley Nautical we're passionate about anything that happens in, on or around the water.

Wiley Nautical used to be called Fernhurst Books and was founded by a national and European sailing champion. Our authors are the leading names in their fields with Olympic gold medals around their necks and thousands of sea miles in their wake. Wiley Nautical is still run by people with a love of sailing, motorboating, surfing, diving, kitesurfing, canal boating and all things aquatic.

Visit us online at **www.wileynautical.com** for offers, videos, podcasts and more.

Contents

Introduction   **1**

CHAPTER 1   What Type of Sailing Do You Want To Do?   3

CHAPTER 2   What Type of Boat Should You Buy?   13

CHAPTER 3   Choosing the Right Hull Material   29

CHAPTER 4   Hull Design   43

CHAPTER 5   Stability   61

CHAPTER 6   Sails and Rigging   73

CHAPTER 7   The Decks   95

CHAPTER 8   Accommodation   109

CHAPTER 9   New or Used?   121

CHAPTER 10   Additional Costs   133

CHAPTER 11   Inspection and Survey   143

CHAPTER 12   Trial Sailing   157

CHAPTER 13   Taking Possession of Your New Yacht   167

CHAPTER 14   Can I Share a Boat?   177

CHAPTER 15   Documentation   185

CHAPTER 16   Buying Abroad   193

CHAPTER 17   Training and Qualifications   201

Glossary   **211**

Introduction

I am assuming that one of the reasons you have bought this book is that you actually like sailing. That may sound a pretty obvious statement, but it is amazing the number of people who buy a boat having never even set foot on one. There is lot to consider. First of all, how is it going to affect your life? Is your partner or your family going to be involved? If not, will they mind that you will spend many hours or whole weekends tinkering about on the boat, bringing lots of dirty equipment home to litter up your dining room? It all looks very tempting on a sunny day seeing a boat bobbing about on the water, but what about days when it is cold, rainy and far from idyllic?

Before you even consider buying a boat, try and experience at least a few hours on the water. If, having considered the above, you and your family are still keen, I hope this book will help guide you on an exciting new venture.

The process of buying a boat can be lengthy, expensive and stressful. Hours can be wasted traipsing around boatyards and marinas looking at boats that are nothing like the owner has described, or presented in such a poor condition that you wonder how well she has been maintained during the previous ownership.

Relying on brokers is not necessarily the answer either, as not all show a great deal of interest in actually finding a boat that fits your wish list. Some do, but often when dealing with smaller yachts, the relatively small commission they make doesn't inspire them with enough enthusiasm to actively search for something to suit your requirements.

This book aims to arm you with much of the information necessary to make a clear judgement on what type of craft would best suit your requirements and advise you on how best to find your perfect boat with the least amount of heartache and expense.

First you need to establish what size and style of boat will meet your needs. The only way to do this properly is to work out what sort of sailing you and your family would like to participate in. You can then start to look for vessels that fit the bill.

This book takes you through every situation you're likely to confront during the process of finding, inspecting, testing, surveying and finally buying the perfect boat for your own personal style of sailing. Subjects include hull and **rig** design, stability, construction materials, accommodation layouts and trial sails. There is also information on legal procedures, licensing, insurance and registration. You will also learn how to buy your dream boat without risking losing your money.

Buying a yacht marks the start of a voyage of discovery. You will never stop learning about yourself, your craft and the countless places to sail, whether local creeks or distant islands. It does not matter whether you buy a modest craft for weekend pottering or a vessel built for circling the world. You will have joined a relatively select group of seafarers who understand the pride that comes with yacht ownership. I trust that this book will help you at the start of that voyage. Happy sailing!

Duncan Kent

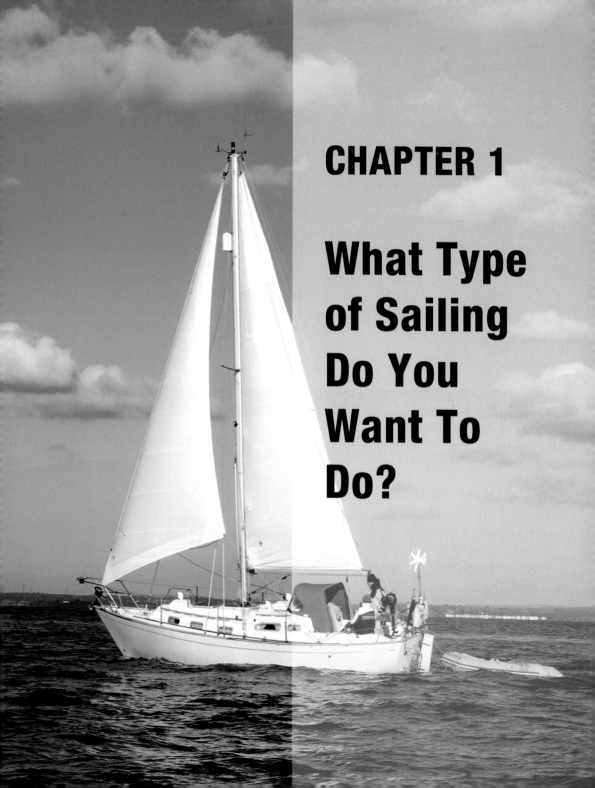

CHAPTER 1

What Type of Sailing Do You Want To Do?

B efore embarking on the long search to find your dream boat, it is imperative to have some idea of the type of sailing you want to do. Without it, you could find you've wasted a great deal of money and time before you finally end up with what you really want – if you ever do. Once you've worked out what you want from sailing it's so much easier to find a boat that will suit you and your family's needs.

Do discuss it at length with your family if they're going to be joining you on the water. After all, if it's to be a family hobby you don't want to end up with a boat that will frighten them all away on their first outing, or you'll be heading for some lonely sailing in the future! Sadly, too many people set out to buy their first boat after reading an inspiring account of cruising in the tropics by some salt-encrusted old gent who sails the oceans single-handed and navigates only by the stars. Whilst this might be the ultimate dream for some, and possibly many of you, you will need to start off with the right boat in order to build you and your family's confidence, before moving on to bigger things. Jumping in at the deep end and trying to learn the basics of sailing in a large, heavy-**displacement**, ocean-going yacht will make the process so much harder – and possibly expensive too, as you gain valuable, but costly experience manoeuvring her in and out of tight marina berths.

Gaining Experience

If you are new to sailing it is advisable to get as much experience of as many aspects of the sport as possible before investing in your own vessel.

Dinghy sailing and clubs

Often the easiest way to get into sailing is by joining a club, of which there are hundreds dotted all around the UK. Don't assume you have to go down to the sea to find one – there are plenty of sailing clubs on lakes and reservoirs that cater for all types of sailing craft from dinghies to quite large cruising yachts.

As an adult, if you want to try your hand at sailing via the dinghy/sailing club route it can be a great way to meet like-minded people and learn the basics in good company. Furthermore, if the club caters for cruising yachts as well as dinghies, you can often try out a few different types of yacht before choosing the one that you feel most closely meets your needs. At the end of the day, the more boats you sail of different types, the better idea you'll get of what sort of sailing you really like doing.

Charter

Chartering a yacht allows you to try before you buy – both the design of the yacht and the experience of living on board. If you don't have the knowledge to sail the boat yourself then opt for a skippered charter. Flotilla holidays can suit experienced dinghy sailors moving up to yachts. There is a wealth of charter opportunities around the UK and abroad.

Sailing school

If you need to learn to sail or improve your skills then taking a course with a recognised sailing school is advisable. When choosing a school, try to find one with a yacht that might be along the lines of what you are after. There is little point learning to crew on a 50 ft race yacht if you are thinking of buying a 28 ft cruiser. The full range of sailing courses is covered in Chapter 17.

Volunteer crew

Skippers are often on the lookout for crew, whether for long passages, summer cruises or regular day-sails. Have a look on the websites and noticeboards of local sailing clubs, or give them a call and ask for their advice. There are even specialist agencies, such as Crewseekers (www.crewseekers.net), that put skippers and crews together. Always establish the financial and insurance arrangements before accepting a crew place.

There are many sailing clubs inland as well

Refining the Wish List

As you gain experience you will develop a better understanding of what sort of sailing appeals to you. As you read through the book there will be dozens of decisions to make as you consider the elements of your perfect boat. To get the ball rolling it may be worth asking yourself which of the following you need your new boat to provide.

Should your new boat provide:

- shelter from the rain
- somewhere to sleep
- somewhere to cook
- a toilet
- an inboard engine
- exciting performance
- a large **cockpit**
- the ability to beach/dry out
- standing headroom
- offshore ability
- separate cabins

You will also need to think about:

- Who will you sail with?
- Where will you keep the boat?
- What is your budget for purchase and annual expenditure?
- How practical are you?
- How many days will you sail each year?

Now think about how each of the following types of sailing fits with your decisions above.

Dinghy sailing

Many people learn to sail in dinghies and there are hundreds of designs from the simple Mirror dinghy to the hi-tech Moth – an exotic-looking, lightweight boat that sails above the water on hydrofoils. What unifies dinghies is the likelihood that you will get wet and the possibility that you will capsize. Dinghies are the cheapest way to get sailing and the smaller ones can be carried on the roof of a car. They can also provide the most fun with some requiring **helm** and crew to dangle acrobatically from trapeze wires as the hull skims over the waves.

Choosing a dinghy is beyond the scope of this book but any sailing club will be able to advise you on the best way to start and the right craft to consider.

Day sailing

If time or resources are limited, or if this is your first boat, then you may want to opt for an open day-sailer that can take four or six people in comfort across relatively protected waters such as estuaries. There will be a large area for seating and this spaciousness is the big advantage

Modern dinghies can be stable, fast and fun

of day-sailers. There may be a small covered area in the **bow** (known as a cuddy) to keep kit out of the rain or spray, but the crew is essentially exposed to the elements. There won't be a toilet or anywhere to cook. The size of the boat and the fact that there is no shelter will limit where and when you sail.

Overnight sailing

One step up from day-sailers are craft that have some form of sleeping accommodation below. The cuddy will be extended into a small cabin though there will not be room to stand up. If you are lucky there will be a chemical **heads** (toilet) and a single gas ring but do not expect a separate loo or **galley** (kitchen). This is more like camping than caravanning and the extra accommodation will eat up space in the cockpit.

This small day-sailer can be rigged and launched in less than an hour

Coastal and inshore cruising

Those of you who are looking to take to the sea with your families and friends will need to be looking for a more substantial boat – one that can handle a wider variety of weather and sea conditions without causing fright or seasickness to her crew. Unlike smaller day-boats, which are often open to the elements, coastal cruising yachts will usually be decked and will most likely have some form of accommodation below, albeit not necessarily luxurious!

The importance of the cabin is to provide some respite for the crew in foul weather, to change clothing or to prepare a cuppa or a meal during a passage. Obviously, the more sophisticated the interior, the more versatile it will be. If you plan to spend nights on board then you'll clearly need bunks and probably a portable toilet at the very least. Water and some means of cooking are useful too, if you don't want to spend money eating ashore.

Of course many boat owners will opt to sail just for the day, then tie up in a marina in the evening to use the showers and to eat ashore, only coming back to the boat to sleep. These boats won't need a sophisticated galley for preparing main meals or a smart heads with a shower. If the price of a pub or restaurant meal is factored into an annual holiday budget, then it might not seem too bad, but if you want to sail regularly at weekends, then the costs of eating out will soon start to mount up.

One of the most common traps to fall into when buying your first cruising boat is to over-specify your requirements. Often prospective boat owners hear stories of others crashing through huge waves and high winds, and feel they will probably need an ocean-going yacht to cope with the occasional spot of stormy weather. But to be honest, with careful planning it is unlikely you'll get caught out in really bad weather when coastal cruising, particularly if you have chosen a route that has plenty of 'fall-back' refuges along the way in case the going gets too tough or uncomfortable.

A small cruiser of less than 30 ft in length can take you almost anywhere

I know of numerous couples of all ages who have sailed on their own for many, many years and always managed to avoid getting into a situation that they can't cope with by dint of cautious planning and meticulous preparation. I have been sailing for over four decades and I can count on the fingers of one hand the number of times I've been caught in a really bad storm whilst coastal cruising. There have been times, particularly in the Mediterranean, when the conditions have cut up rough and we've had to run for the nearest bolthole. But before I set out I always read up on the possible fall-back ports or sheltered anchorages along my route and note down a few important navigation marks and transits, so that, if it should by any chance look as if it might be about to turn a bit nasty, I already know where I can go and what to expect when I get there.

So, if your ideal is to stay in sight of land most of the time, then you'll only waste time and money looking for a boat that is designed to ride out storm-force winds in the middle of an open ocean. After all, if you fall in love with sailing so much that you decide you want to start going further afield and crossing large oceans, you can always upgrade to another boat that is specifically designed for offshore/ocean use, by which time you'll probably have a much better idea of what you really want anyway.

For those who just fancy a spot of fair-weather cruising within sight of land and then stopping in a marina or sheltered anchorage for the night and using shore-based facilities, a modern, spacious design with high interior volume will be more suitable and child-friendly and you'll have much more space to entertain guests – maybe even accommodate them for the occasional night.

As soon as you've some idea of what boat you think might suit the style of sailing you're interested in doing, then I highly recommend you charter an identical, or at least very similar yacht to get a real, hands-on feel for her.

Offshore and ocean cruising

There are some folk who suddenly decide, out of the blue, that the time has arrived for them to embark on a life-changing expedition. These intrepid adventurers often go straight out and buy a blue water cruising yacht before they really know what they need, or even want. If you have pots of money and are happy to heed the advice of someone well experienced in this type of venture, then you might just get away with it. But it'll cost you. By far the best way is to get used to sailing in a smaller coastal cruiser first, which will give you many more ideas about which boat designs and equipment you would really want for long-term sailing and living aboard. It also means that when you make mistakes, which you undoubtedly will, the inevitable reparations won't totally clean out your bank account!

Open-ocean sailing in a monohull requires a sturdy, deep-keeled vessel with a sea-kindly motion. The latter is more important than some people realise when choosing an ocean cruiser. It is vital that you can live, eat, sleep and work aboard your boat in mid-ocean, day and night. A yacht that slams into an oncoming sea, or is too heavy for the autopilot to steer in high winds, can be a nightmare on prolonged passages – tiring the crew and seriously increasing the likelihood of accidents caused by careless mistakes.

If you plan to live aboard for extended periods, you have to consider the accommodation and the yacht's systems extremely carefully. If you have little experience of cruising, then unless you're fortunate enough to have friends who have already done it, and you're willing to take someone else's word for it, you're unlikely to be able to choose the right yacht for the job or prepare it properly for ocean sailing.

Before you make up your mind and start throwing your money around ask yourself a few soul-searching questions. What would I/we do in a life-threatening emergency, such as a man overboard, thousands of miles offshore in the middle of the night? What if the diesel tanks become infected with the fuel bug so the engine fails and there is no means to charge the batteries? Or the water-maker dies? Or the sails get destroyed in a storm?

These are just a few of the more serious, but nevertheless not untypical incidents that can happen at any moment when cruising across oceans. And as important as good quality equipment might be, the ultimate key to your survival is within yourselves. It's easy to be full of bravado before you leave your home country for months, possibly years, but are you sure you're 100% ready to leave it all behind and become full-time seafarers?

Some people jokingly state that before you think about going blue water sailing you should try living in a cupboard for a week while someone rocks it back and forth violently and then opens the door to shower you with cold water every couple of hours! Okay, so this might seem overly pessimistic, but I wouldn't want to encourage someone to go to sea who doesn't know exactly what they are likely to face, and is not fully capable of coping with the worst that might

A more seriously equipped yacht is required for ocean cruising

happen – especially when they're involving their partner and/or family in such a massive lifestyle change.

That said – hundreds do it and most love every minute of it and only wonder why they didn't do it many years earlier. Many take children, who experience things first-hand that can only be described in the classroom. It can be a fantastic way of life, a great way to see the world and make new friends, but it is not for everybody.

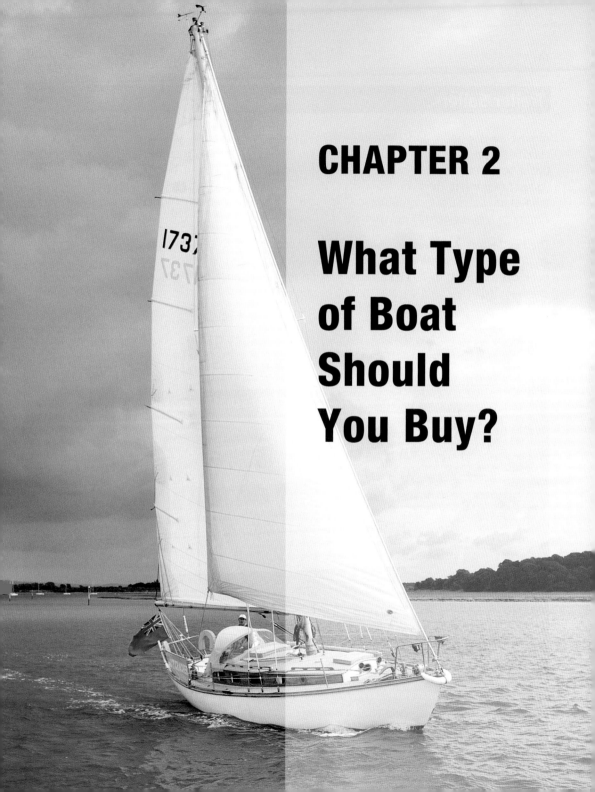

CHAPTER 2

What Type of Boat Should You Buy?

Trailer-Sailers

Almost all day-boats and most overnight-sailers can be towed behind a car and so can be classified as trailer-sailers. Most could also be left on a swinging mooring or in a marina berth, but if you can simply hitch the boat up to your car and take it back home with you when you're not using it, then you'll avoid shelling out for storage and/or mooring charges – often the biggest annual expense for a boat owner – and have more spending money to keep her in tip-top condition and replace old equipment.

The key factors you need to bear in mind when looking to buy a trailer-sailer are size, weight and the ease and speed with which she can be rigged and launched. Lightweight dinghies and open day-boats up to 750 kg and 16 ft in length rarely cause problems with towing and can be manhandled easily on a slipway by one or two people.

Trailer-sailers usually have retractable keels for ease of towing

Rigs on this size of boat are usually designed to be raised by hand – either by brute force, or by the use of some sort of A-frame tackle. In the latter case these mechanisms are usually provided along with the boat. Some even remain attached to the boat permanently.

Genuine trailer-sailers will most likely have some form of retractable keel as well. Although it is possible to trail a bilge-keeled boat, they are usually a little more awkward to launch and recover given the depth of water required for them to float on and off the trailer. For this reason, with a bilge-keeler you really need a separate launching trailer to keep the road trailer's wheels, bearings and brakes from being submerged in saltwater and subsequently damaged.

With a retractable keel – whether it lifts or swings up – it will nearly always take up some of the accommodation space below deck in the form of a keel box. On smaller boats this can be a little restrictive, but some more clever designs have this built into the furniture, so it's less intrusive. Retracting keels are usually raised by some form of mechanical **winch** (see Chapter 4) and often it's preferable if this is on deck rather than down below – especially in a small boat without a great deal of winch-handle swinging room below.

Inside the Anderson 22 the keel box becomes part of the furniture

Most trailer-able boats are powered by small outboards, which are easy to take home for safe keeping and servicing. Nowadays the smaller, lighter two-stroke outboards (those that run on an oil/petrol mixture) have been effectively outlawed due to their inability to meet the latest stringent EU emission levels. Use of the old engines is still legal, however, so you might well find you inherit one with the boat. Two-stroke engines with the more sophisticated oil-injection systems are the less polluting of these motors, but older ones, such as the Seagull (where you have to mix petrol and oil together in a container before fuelling it up) are most definitely becoming collector's items.

All new outboard motors use the more complex four-stroke petrol engine. They are often quieter and less smoky than their two-stroke counterparts, but their substantial extra weight needs to be considered, particularly when towing and manhandling it on and off the boat. Four-stroke outboards are also fussy about the way you lay them down. Ideally they prefer to remain upright all the time, as this keeps the engine oil where it should be – in the sump beneath the engine. But there is usually a prescribed way to lay them down horizontally during transportation that doesn't allow the oil to back-feed into the cylinders. If you get it wrong and the engine cylinders fill up with oil, you not only risk it seizing up, but also causing possible damage to the valves – that is, if you manage to get it to fire up at all.

The Hurley 20 is a surprisingly roomy boat down below

If you choose a boat over 750 kg or 16 ft length overall (**LOA**), you are starting to look at towing a much larger (possibly four-wheel) trailer, requiring a more powerful towing vehicle and a few extra crew to help you rig and launch her. In return, though, you'll have a boat that you can live aboard in reasonable comfort for long weekends, or even the odd week or two without too much discomfort. I lived happily on a 20 ft Hurley bilge-keeler in the past and spent several nights with another two burly fellas on board as well. Apart from the tumultuous snoring and the game of twister in the morning as we scrambled for our wet-weather gear, we coped perfectly well – although the heads were strictly out of bounds!

The right vehicle

Of course there are going to be limitations on the size of boat you can tow on a trailer and this is usually dictated by the size of your car. Pretty much the largest or heaviest boat you can tow with a normal car in the UK is around two tonnes (2 000 kg). Any heavier than this and you need a specialist vehicle, which will once again start racking up the costs, unless you simply hire one occasionally when you want to move it. Realistically, most medium-sized saloon or estate cars can safely tow up to a tonne (1 000 kg) or so. Larger, long-wheelbase estate models may do so up to 1 500 kg, but it is generally recommended that you shouldn't tow more than 85% of the towing vehicle's kerb weight (vehicle weight plus 75 kg per person and 1 kg per litre of fuel carried). There are strict regulations governing towing in the UK and Europe, a comprehensive summary of which can be found at the UK government's website (www.dft.gov.uk). What's more, it's not just the fully-loaded trailer's weight you have to worry about – the total combined weight of vehicle and trailer is limited as well, usually by the car's manufacturer, so don't think you can necessarily pile an outboard motor and all your gear into the boot without doing a few sums first!

All modern passenger-carrying vehicles have a recommended maximum towing load weight stated in their handbook and on their vehicle identification number (VIN) plate, so that the optimum braking performance of the vehicle isn't compromised. If in doubt, contact the maker of your car or check out the towing specifications on the Internet.

Another factor that restricts what boat you can trail is its physical size. The width and length of the vehicle and trailer together must not exceed certain rules laid down by law. Generally, if a vehicle and fully loaded trailer's combined gross weight do not exceed 3 500 kg, then you are limited to within 2.55 m width by 7.00 m in length for the trailer and its load.

It is also worth noting that trailers capable of carrying a load greater than 750 kg must have brakes that automatically activate when the towing vehicle itself applies its brakes and a safety chain to keep the trailer attached to the car, should the tow-ball mechanism fail. Trailers over 1 500 kg must have a device that will also engage the trailer's own braking system should it break away from the towing vehicle.

Get some local advice before you launch into unknown waters

Apart from having to ensure that your vehicle and trailer are a safe combination to drive on public roads, it is easy to be put off trailer-sailing by worries about actually launching the boat.

Providing you heed the advice from those who know and understand the pitfalls – particularly locals and regulars to the particular launch slip that you plan to use – launching a boat is not at all problematic or dangerous. As with most things concerned with boats and boating, it's all in the planning – well, maybe not *all*, but knowing what to expect and preparing yourself for a variety of possible problems can only stand you in good stead should the worst happen.

Large trailer-sailers

A number of owners buy 'trailer-able' boats, but not necessarily trailer-sailers. By 'trailer-able' I mean boats that are within the limit for towing on a double-axle braked trailer behind a powerful 4 × 4, but are then rigged and launched for the entire summer season – leaving them on a mooring or marina berth when not in use.

The 24 ft Cornish Crabber is an ideal size for towing

This can be an economic way of sailing, provided you have somewhere cheap or free to store her for the rest of the year. As with all trailer-able boats you also gain the advantage of being able to take her to a new cruising area each year, even abroad, without the trouble of having to sail there first.

Probably the largest boat you could self-trail would be around 28 ft long, but that would also depend on its weight. Some power-sailers (a cross between a sailing yacht and motorboat), such as the MacGregor 26, have water ballast that can be emptied out on the slipway before you tow them away. This considerably reduces her overall towing weight, but has the drawback of not being as effective as iron or lead ballast when it comes to keeping the boat upright under sail.

In my experience the ideal cruising trailer-sailer should be no more than 24 ft long and 1.5 tonnes (1 500 kg) dry weight all up – something like the Cornish Crabber 24, for instance. This would require a double-axle, fully braked trailer and probably a 4 × 4 or at least a large two-litre car to tow her safely. Even then you wouldn't want to be putting her in and taking her out every day!

A good number of marinas now 'dry stack' boats on tall racks in their yards, getting them down with huge forklift trucks and launching them in advance of you going sailing. Although this is primarily for rigid inflatable boats (**RIBs**) and small sportsboats, this could be perfectly viable for a trailer-sailer with the rig down and strapped over the **coachroof**. Dry stacking is a much cheaper way of storing your boat and, if you don't leave it in the water for too long without moving, probably eliminates the need for any antifouling as well.

Inshore/Coastal Cruisers

To my mind, branding a yacht as an 'inshore' or 'coastal' cruiser is some-what misguided, but unfortunately these expressions are commonly used in the industry. As far as I'm concerned a sailing yacht that is designed to go to sea in tidal waters is just that – a yacht.

With the mast down and keel up you might be able to dry stack a trailer-sailer

So what makes an ideal sailing yacht for coastal cruising? Obviously when cruising within sight of land and stopping overnight at anchor, on a visitor's buoy or marina berth it isn't necessary to have a boat that will withstand storm-force conditions. This is simply because it will be quite rare for you to experience such weather when keeping within sight of land and within earshot of a reliable weather forecast. Some of you might well prefer to buy an ocean-going yacht even for coastal cruising, as a 'belt and braces' precaution, and I wouldn't want to deter you from doing so, but in general, prospective yacht owners need to know that it's perfectly safe to go coastal pottering in something less robust than a hurricane-proof ocean cruiser.

A lightweight, fast performance yacht can get you home quickly if needed

Some experienced yachties swear by lightweight, fast, performance boats for coastal sailing. The theory is that if the yacht is fast and can still be safely sailed in winds up to and including Beaufort Force 7, then you're more likely to reach shelter before the worst of any weather system can hit you. In certain ways there is good logic to this, providing you're not caught out in an unpredictable storm one day with a rather flighty, lightweight racer/cruiser in which to do battle with big seas.

This suggestion, though, goes against a well-established golden rule in sailing – if the weather blows up head out to sea, giving yourself enough sea room to analyse and fix any problems. In many cases the closer you are to the coast in stormy weather, the greater your chances of running into trouble, so a well-found yacht will make this option a safe and comfortable one.

Staying out during a storm is, of course, contrary to one's natural instinct for survival – to sail as fast as possible for the nearest safe haven – and it often requires considerable determination not to choose what looks like the easy way out. But unless your haven is a large, well sheltered harbour with an entrance that is safe to negotiate under sail in all conditions, then you're better off waiting it out well offshore where the waves are usually less steep and you don't have to worry about rocks or other vessels heading into port.

My ideal coastal cruising yacht is a compromise between a reasonably fast, fun-to-sail day-sailer and a boat that can withstand the occasional blow and 2 m high waves without frightening or throwing my crew around – particularly if I have my grandchildren on board. There are quite a few of these boats around and for a decent price as well, if you're prepared to do a little work on them.

Offshore/Ocean Yachts

Running for a safe harbour is not always the right answer

I would class an offshore-rated yacht as one that has been designed to take care of its crew in all open-ocean conditions – right up to hurricane-force winds and 10 m high seas. This is sometimes overlooked when looking for a boat in which to go blue water cruising, but is very high on, if not at the top of, my list of priorities when choosing a yacht for ocean passages. Naturally, it is equally vital that the yacht is seaworthy, strong and safe, but if it has a violently uncomfortable motion at sea, then it isn't, to my mind, a true offshore yacht.

What I mean is that the vessel should exhibit a comfortable, controlled and well-balanced motion at sea, so that the crew are able to sail, cook, eat and sleep easily without suffering too much discomfort. Second only to the yacht being well found, the crew is what makes a long-distance cruiser a safe and secure means of getting from A to B. A tired and irritable crew is one that will start making mistakes when the going gets tough. The better they are fed and able to make the most of their off-watch time, the happier and more conscientious they will be, so that if their skills are called upon for a sail change during a 40-**knot** squall at 3 am, they'll be able to take it in their stride and not risk the lives of themselves or their fellow crewmates by being exhausted and inattentive.

So, what exactly makes a yacht sea-kindly? Well first and foremost is its motion through or over the waves. A lot is said about modern, lightweight glass reinforced plastic (**GRP**, or glass-fibre) yachts 'slamming' into oncoming waves rather than slicing through them. This slamming not only shakes the hull and rig but also jars the crew's nerves and throws everything out of the lockers below deck. Consequently, it doesn't just test the integrity of the yacht to its limits, causing excessive wear and tear to the rig and fittings, but it also drags the crew's morale down and stops them being able to sleep, cook, ablute or simply relax when they're off watch.

As with most aspects of sailing, there are several schools of thought with off-shore yacht design, but it is generally accepted that ocean-crossing yachts should be of a higher displacement

This Contessa was designed and built to take heavy weather at sea

than coastal cruisers and that they should have a deeper, vee-shaped forefoot (the forward section of the hull that meets the waves first) to enable the hull to slice through the waves. The deep-vee dampens the hull's descent into a following wave, gradually increasing the water resistance the deeper it penetrates. Clearly you don't want too much of a 'nodding dog' effect, as this will slow your progress and can cause seasickness among the crew. One also has to be a little careful not to place too much load forward, which will make her bows dig too deep into the oncoming wave, causing her to scoop up copious quantities of water.

There is a school of thought that says 'the quicker you can go, the quicker you're out of it', encouraged by those who believe that a faster, light-displacement, flatter-sectioned hull will allow you to sail around and avoid the worst areas of bad weather within a front or weather system. Well, this might be true of an Open 60 class racing thoroughbred averaging 20 knots

Fast yachts might get you home fast, but they can offer a rough ride!

or more, but with most production cruising yachts up to 50 ft LOA, the result is usually a very uncomfortable, if somewhat quicker and wetter, ride through the storm.

Blue Water Cruising Yachts

Blue water cruising is a term that conjures up thoughts of endless lazy days flitting from one tropical-island paradise to the next in a perfect, warm 15-knot breeze (something that is in fact fairly rare I hasten to add).

Actually, a blue water cruiser is simply an offshore/ocean cruising yacht that has provision for living on board for extended periods of time under a wide variety of different circumstances. Usually they will be better equipped with items like water makers, generators, freezers, solar panels, wind generators, laptops, satellite phones etc., but the style and design of the yacht itself will mostly be identical to an offshore/ocean-class yacht.

Long-term blue water cruisers often start off with a plethora of fancy gizmos to cater for their every need whilst living aboard. I've seen boats leave the UK weighed down with bread makers, hair dryers, curling tongs, toasters, electric kettles etc., all powered by a truck-load of golf-cart batteries hooked up to a huge inverter. But by the time they've reached the Caribbean with a few thousand sea miles under their belts, there's no longer any sign of this frivolous domestic paraphernalia.

'Oh, we soon got rid of that lot,' they inevitably say. 'We quickly became fed up with the noise and heat of perpetually running the engine to charge the batteries – and it costs a lot on fuel too. Besides, most of the fancy domestic gear rusted up solid in the first few months and we couldn't afford to replace it anyway.'

The trick with blue water cruising is to abide by the well-established 'KISS' principle espoused by nearly all experienced liveaboards – Keep It Simple Stupid! The less there is to go wrong, the less time and money you'll spend trying to get it all fixed and the more time you'll have to enjoy yourself and the sailing.

I've seen plenty of other examples, most notably at the start of the Atlantic Rally for Cruisers (ARC) that is run every year from Las Palmas in the Canary Islands to St Lucia in the Caribbean. Skippers, about to make their first Atlantic crossing, panic at the last moment because their fancy new colour **chart plotter**-cum-radar-cum-depth-sounder display isn't working perfectly. But what are they going to do when it stops working a thousand miles from the nearest land? Often it does, and more often than not the skipper arrives in Rodney Bay, St Lucia bang on time with a big grin on his or her face saying, 'Yeah, the plotter died after a few days, along with the water maker and we had to stop using the electric kettle because we got fed up with the noise of the engine running for four hours a day just to charge the batteries. But we managed fine – it was kinda' fun getting the sextant out and plotting our own fixes and water wasn't a problem once we'd rigged the bimini and sail cover to collect the rainwater during the frequent squalls.' Needs must, as the saying goes, and blue water cruisers very soon discover better/ simpler/cheaper ways of doing things that they'd never envisaged before they set off.

The question all potential blue water cruisers should be asking themselves, both when they're deciding on which yacht to buy and when they're planning the equipment inventory, is 'Do we really need it? Isn't there another, simpler way we can heat water, make power, navigate etc.?' More often than not there will be an alternative method of doing something that will be easier, cheaper and most likely a good deal more environmentally friendly than going down the high-tech route. At the very least, you should be prepared and equipped to instigate at least one alternative should your clever gizmos refuse to co-operate halfway across an ocean.

Sometimes you can simply have too much gear

Monohull or Multihull?

Whether you want to buy a dinghy or an ocean cruiser you first need to decide on a few basics. The majority of new boat-buyers tend automatically to start looking for a monohull, with only a few giving multihulls a second thought. However, it's well worth considering a yacht with more than one hull. Do try a few catamarans, or even trimarans out – before dismissing the idea out of hand. You might find the level sailing (multihulls heel much less than monohulls), greater deck space and higher speeds to be well worth the disadvantage of having a larger boat to park and slightly reduced load-carrying ability.

Small catamarans, such as the famous Hobie Cat, can be fantastic fun for all the family, although they do take a bit of getting used to and you have to be prepared to get wet. Small

Catamarans can float in very little water and take the ground easily

trimarans also make great day-boats and most are designed to fold up and fit on a trailer up to 16 ft or so.

For coastal cruising there are certainly more monohulls available, although over recent years cruising catamarans have become decidedly more popular thanks to their extra deck and cockpit space, which is a real boon when entertaining guests or taking a gaggle of hyperactive young kids for a day out messing around on the water. Of course you can do the same in a monohull, but a cat always seems to have so much more deck space for dripping bodies, surfboards, kayaks, a pile of inflatable dinosaurs and whatever else kids like to play on in the water!

Catamarans also have a shallow draught, which means they will float in just a foot or two of water – so you can get right in close to the beach to anchor, or creep up a shallow creek where the majority of monohulls can't go. They will dry out easily too, so if you fancy parking up on the beach for the day, either to barbeque or to give the hulls a quick scrub, it's not a problem.

CHAPTER 3

Choosing the Right Hull Material

Wooden Hulls

Traditionally-built wooden boats exude character and are invariably a delight to look at when well maintained, but they can try the patience of even the most dedicated boat lover and can relieve you of all your spare time (and cash for that matter) if you pick a duff one. A properly looked-after and greatly loved classic wooden boat can be as good an investment as a GRP or metal one, providing she has been correctly maintained throughout her lifetime and hasn't been 'got at' by a less-than-knowledgeable DIY enthusiast.

These days the few boats that are built entirely from wood often come sheathed in an epoxy coating to pro-tect the wood from the effects of ultraviolet (UV) sunlight, and to make it completely impervious to water. But such modern techniques are miles away from the principles established over the centuries when wood really was the only material with which to build a boat. Traditional wooden hulls

Traditional wooden yachts can be a delight to own if you have time to spare

were built around a sturdy solid wood framework using a strip planking method or, after marine plywood had been invented, by forming sheets of ply around the frames – a process called cold moulding.

Some planked boats were built in a 'carvel' fashion, whereby each plank simply butted up against the next. The wedge-shaped gaps between planks were plugged, or caulked with a fibrous string called caulking cotton. This 'caulking' was wedged between the planking joints with a mallet and a caulking iron to form a watertight seal. When the boat was first launched she often took on a little water for a few days, until the water swelled the caulking sufficiently to completely seal every joint.

Other wooden boats were built using the **clinker** principle, whereby each plank is laid so that it overlaps, rather than butts up against, the adjacent lower plank. The ribs and frames used are

A carvel-built hull is smooth and rounded, but needs regular maintenance

the same as in a carvel boat, however clinker planking is usually slightly thinner, so the hulls end up somewhat lighter.

Another popular and comparatively cheap method of building in wood is to use a special type of plywood called Bruynzeel marine plywood, which utilises a powerful waterproof glue to hold the veneers together. Plywood hulls can be constructed using a few different methods. 'Hard **chine**' hulls have plywood panels fixed onto frames as flat sheets. This technique is also used for some metal hulls as it saves the complication of bending the material to suit the required curves. Enthusiasts often praise the hard-chine building method because providing the designer gets his lines correct, the chine creates a flat spot on which the hull sits at the optimum level of heel and provides an increased resistance to it heeling over any further.

The planks overlap each other in lighter, clinker-built hulls

Other ways to build in ply include the **cold-moulded** method, whereby a flat sheet of plywood is simply bent over a series of frames and fastened down. Being quite flexible, plywood takes fairly easily to this method of build and the wood eventually settles into the shape naturally, although it does put more strain on the fastenings initially. Often you'll hear the expression 'diagonal planking' – this is where a second (possibly even a third) layer of plywood is fastened over the first with the sheets placed diagonally opposite to the layer below, effectively doubling (or tripling) the strength of the hull and covering over the inner seams.

The advantage of plywood is that it's easy to use for amateur boatbuilders and it creates a very strong monocoque construction without relying too much on the inner framework for overall strength and integrity. This is because plywood is equally as strong whichever direction you apply

Plywood is a very popular material for boats, which often have chines

a load, whereas planking is inherently weaker across its grain and needs strong frames to support it in that direction.

If the bare plywood is then sheathed with epoxy, it behaves in a similar fashion to an all-GRP hull and is almost as water-resistant. However, if a plywood hull is damaged or punctured and a repair isn't carried out almost immediately, the unsheathed wood will behave like blotting paper and absorb the water into its veneer construction, quickly softening the glue and rotting the wood.

When looking at the possibility of buying a wooden boat it makes sense to speak to someone who knows about them, and preferably someone who owns or used to own one of the same construction for a few years. They will have discovered the boat's weak areas over time and can show you where to look out for problems. While this won't do away with the eventual need for an experienced surveyor, it should allow you to eliminate a few no-hopers before you start spending money on professionals.

A serious refit job is going to be required here!

Unfortunately, the most vulnerable areas to look out for are usually the hard-to-reach places, which require just as much maintenance as the more accessible areas. Deck-shelf beams are a common area of rot, as are the joints where the coachroof is fixed to the deck. In particular, look out for upright planking where the vulnerable ends might just have had a piece of trim keeping the water out of the wood grain. In fact wherever there's a piece of added trim you're more likely to find problems with water ingress and/or rot.

As with all types of yacht, but particularly with wooden boats, it pays to find an owners' association for the boat you are interested in. Here you can pick up myriad valuable tips and hints about what to look out for and what to be wary of as you look over the boat. Before making an offer on your future dream boat, do get a specialist wooden-boat surveyor to check it out, or you might well find yourself with a serious and expensive reconstruction job on your hands.

Metal Hulls

Quite a few yachts, particularly larger offshore cruising yachts and specialist boats such as work-boats and safety tenders, are fabricated from metal – usually either steel or aluminium. There are one or two mid-size production aluminium yachts being built, which, despite their higher initial purchase price, have proven very popular among the cruising fraternity for a variety of reasons. Yachts like the Ovni range, made by the French company Alubat, are probably the most commonly seen alloy yachts in Europe. The entire range has lifting keels and rudders, hard-chine hulls and largely unpainted topsides, which certainly makes them stand out in a crowded anchorage – apart from the fact that they're probably also the ones anchored in knee-deep water, or possibly even parked up on the beach!

Aluminium is a superb boatbuilding material because it is strong yet lightweight. There are downsides, of course: it's difficult to get paint to stick, repairs require specialist welding and

The aluminium-hulled Ovni has hard chines and a retractable keel

it's extremely vulnerable to **galvanic corrosion**. This problem arises when whichever metal is less cathodic or 'noble' than the other corrodes and disintegrates. But aluminium is often used for custom yachts because the strength and integrity of a metal hull doesn't rely as much on frames, stringers or **bulkheads** as wooden and GRP designs do, which means the designer has much more flexibility.

Steel was a very popular material for long-term cruisers from the 1970s onwards as it is extremely robust and can undergo welded modifications or repairs almost anywhere in the world where there is commercial industry. Many have been self-built from kits or plans, with some of the most prevalent ranges being from the board of Robert Tucker, or the Bruce Roberts design team. Of course steel yachts are very much heavier than their counterparts in wood, GRP or aluminium,

A typical home-built Robert Tucker with a steel hull and wooden coachroof

which makes them rather ponderous under sail, but many long-distance cruising yachtsmen are happy to sacrifice speed for comfort of motion, and the innate resilience to sharp rocks or uncharted Pacific-island coral heads.

The key to a good steel boat is the initial inner hull treatment and the continued outer hull and superstructure maintenance. Clearly, steel rusts very quickly in the open air and even quicker in salty air. The steel plates must be completely clean and free of oxides and weld carbons before any primer is applied and only special paints should be used if the hull is to stay rust-free for much of its life. It's safe to assume that some amateur builders won't have been able to prepare and paint their home-built steel hulls to the required standard, and that the initial coating will not have formed an immediate impenetrable seal. As a friend of mine who once owned a well-built steel yacht quite rightly said, 'Rust never sleeps'.

After priming, a metal hull needs insulation – against both sound and temperature. A large metal hull will reverberate like a huge bell if it isn't suitably insulated inside, with every thrum of the engine or clang of a **halyard** amplified throughout the boat – not conducive to a good night's sleep. More importantly, a bare steel inner hull will create a considerable temperature difference in certain climates, which will cause copious condensation. Both of these factors will make living on board extremely unpleasant, as well as creating corrosion problems where the condensation collects in inaccessible corners.

Another thing to be wary of on metal hulls is any area where two different metals touch. Differing metals, when in electrical contact, and submerged in an electrolyte such as saltwater, will create galvanic corrosion. This can happen extremely quickly, so one needs to keep a regular eye out for any signs. One of the scariest scenarios can occur when a bronze **skin fitting** has been fitted to a steel hull and the insulating seal has worn, thus allowing the two metals to touch. In certain cases the bronze fitting could corrode and possibly even fall out! The same result is achieved when someone ignorant of this process thinks it a good idea to electrically bond everything metal together with a big earth cable. Whilst this might be considered okay for wooden or GRP yachts, it can have lethal consequences in metal-hulled boats.

As with many plywood hulls, a large number of steel yachts will have been built using the hard-chine method, as it is by far the easiest way of constructing the hull. This is often a first clue as to whether the boat has been self-built by an amateur, although not always clear proof. It is quite possible to build a steel boat using pre-formed sheets of steel that have been repeatedly rolled by an enormous machine to form a curved shape that follows the designer's hull lines. This is a more expensive way of building a steel boat, but the end result does look considerably more professional and often has far fewer welds (possible weak points) to worry about.

One last piece of advice from my Bruce Roberts-owning friend, 'A rusty-looking steel yacht can often be a bargain as it will nearly always look in a worse condition than it really is'.

The battle against rust on a steel hull is never-ending

Glass Reinforced Plastic (GRP)

GRP, or glassfibre (fiberglass in the US) as it is commonly known in the marine trade, has evolved significantly over the 50 years or so in which it has been used for boatbuilding. It is now far lighter, stronger and more durable than it once was. Not much was known about the strength and resilience of GRP in the 1960s when it was first used, so early GRP hulls tended to be laid up very thickly indeed. Whilst this appeared to be desirable, even commendable in sheer strength terms, it also meant the boats were very heavy and consequently quite sluggish under sail.

Early boats almost always used simple Chopped Strand Mat (**CSM**) – a glassfibre cloth where the fibres have been chopped up into small lengths and bonded together in the form of matting – and were laid up by hand. Firstly the **gelcoat** (a thin, smooth outer coating) is sprayed onto the inside of the waxed female mould, then the first layer of CSM is placed inside the mould and 'wetted out' by the laminators using resin and rollers. Once the layer has cured enough to become tacky, a second layer is added . . . and so on, until the required thickness is achieved.

A laminator rolling out the resin-soaked matting on a GRP deck

This process is still used in many boatyards to this day, although the materials have improved enormously. What often happens now is that the resin is applied first and then the CSM is sprayed onto it using a CSM gun, which chops it up as it goes. It still has to be rolled out afterwards, however, to ensure the CSM is well soaked in resin and that there are no air pockets in the lay-up.

Developments made in the 1990s and this millennium have meant that hulls can be laid up a great deal thinner without losing any of the strength and durability of heavier boats. This is because builders often use unidirectional matting between layers of CSM to build in extra strength. In addition, strands of tough new materials such as Kevlar and carbon are now interweaved with the unidirectional glassfibre matting, greatly increasing the stiffness and damage-resistance of the final composite.

A process was introduced around the turn of the millennium called vacuum infusion technology and is now commonly used for guaranteeing the integrity of a GRP moulding. Vacuum infusion involves laying up all the cloth and reinforcing fibres dry, enclosing them all in a sealed plastic bag and sucking out all the air out. A pre-calculated quantity of resin is then injected through numerous valves in the bag forcing the resin to penetrate every millimetre of the lay-up at a controlled and even thickness. This process guarantees the uniformity of the resulting composite and the amount of expensive resin used can be strictly controlled, so there is no waste. Also, the toxic solvent fumes created during the curing process are contained within the bag and can be disposed of safely.

The resins used in glassfibre moulding these days have also improved significantly, with most builders now using the more water-resistant polyester products to ensure that their lay-ups remain watertight. Some boatbuilders go one step further and use epoxy resin in their hull lay-ups. This is almost completely waterproof and very strong when combined with high-tech fibre materials. This method is primarily used in racing yachts due to the lightness of the structure, as well as the higher cost. Epoxy resin can also be coated over bare gelcoat as an added protection against **osmosis** (water ingress into the GRP lay-up) and is commonly used to coat a hull that has had treatment to eradicate existing osmosis. Epoxy coating may be offered by a boatbuilder from new, as an extra protection against osmosis. But, in my opinion, if the builder has used a polyester gelcoat, there will be little need to add an epoxy coating as well.

Osmosis is not uncommon in GRP hulls built from the 1970s through to the 1990s, but is almost unheard of in today's mainly polyester resin composite hulls. You will undoubtedly come across it during your search for a yacht if you are looking at boats built prior to the year 2000. Osmosis is often not as bad as it is made out to be and many affected boats go on to last decades after remedial treatment has been carried out. This treatment usually involves stripping or planing off the gelcoat and top layer of waterlogged GRP, washing it regularly with freshwater and then force-drying the hull. Sometimes large vacuum pads are put on the hull as well to literally suck out the moisture.

You won't need to know all the technical background to the chemical process of osmosis when you're buying a boat, but you will need to know how to recognise the symptoms. There are, of course, different levels of osmosis and it is very common in some particular makes of boat. It was made worse during the oil crisis in the early 1970s, when resin prices went through the roof forcing boatbuilders to try and economise. Needless to say this left the GRP lay-up full of air pockets, which, due to the more porous gelcoats used then, allowed large quantities of seawater to seep through into the GRP composite. Osmosis presents as one or more gelcoat blisters, ranging from the size of a pinhead to the width of a full hand-span. Surveyors will puncture these blisters and test the contents of the fluid inside using litmus paper to confirm whether it has indeed been caused by osmosis, or whether the layers of GRP have started to delaminate (peel away from each other) for some other reason.

Osmosis, or boat pox as it has become known colloquially, is rarely fatal for a yacht and I know of none that have been lost at sea because of it. However, something obviously needs to be done

Treating an osmosis-ridden hull using vacuum bags to suck out the moisture

to cure it once it gets too bad, or else the hull could begin to delaminate. If a boat that you're interested in has recently been treated for osmosis, ask to see the guarantee (usually five years minimum) and check out the shipwright who did the work to see if they have a respected standing in the marine GRP repair trade. If all seems in order, then there should be nothing to worry about. If, however, you are not offered any paperwork to show when or where the work was carried out, then I would be somewhat wary about making an offer. If you do, make sure that it is conditional on a surveyor thoroughly checking out the hull to his and your satisfaction.

Some yachts on the market will be showing signs of osmosis, but the vendor will most likely have reduced the price to take account of the cost of the remedial work. If this is the case, then you still need to get a surveyor and preferably an osmosis treatment expert to check it out closely. Get them to confirm the cost of having it properly treated, before deciding whether the asking price is a fair one or whether you should make a lower offer.

CHAPTER 4

Hull Design

The last 100 years has seen significant change to the design of a sailing yacht's hull. The emphasis today is predominantly on combining sailing performance – that is, speed and agility – with a spacious, bright and luxurious interior, whereas in the middle of the last century the most important element of a yacht's design would have been its seaworthiness. A cruising yacht had to look after its crew regardless of the conditions at sea and a keen performance under sail was a nice bonus if you could get it as well. Of course, there were plenty of racing yachts designed for speed, but few would have ever been taken by a family for a week's holiday, or used to cross an ocean for pleasure.

Leaving aside the particular fans of 100-year-old racing yachts or of classic wooden cruisers, the majority of people looking to buy a boat will likely be searching for something built from the 1960s onwards. In the '60s, classic cruising yachts such as the Nicholsons, Elizabethans and Pioneers were in vogue. With their conservative rigs, long, deep keels and high ballast ratios, these yachts would look after their crews in heavy seas, but weren't particularly fast around the cans.

During the 1970s, however, the international racing rules started to have a considerable influence on yacht design – particularly in the shape of the hull. Technically, the maximum speed a displacement yacht could attain through flat calm water used to be calculated by the following formula: maximum hull speed = $1.34 \times \sqrt{\textbf{LWL}}$ (waterline length in feet). So, for rating purposes, the trick was to keep your static waterline length to a minimum, but to design the hull so that once it was heeled, the waterline length would increase as much as possible. Hence the International Offshore Rule (IOR) racing rules in the late 1970s created a stream of yachts with **beamy** middles, 'pinched' ends and generous bow and **stern** overhangs. While this was fine in terms of sheer performance, it made the accommodation below very cramped, the foredeck was usually small and tended to duck its occupants under the waves in choppy seas and the side decks would disappear on reaching the cockpit **coaming**, making it tricky and somewhat dangerous to go forward in a seaway.

A typical IOR-style yacht with wide beam and pinched ends

Since those days cruising yachts have grown ever wider and now most retain virtually maximum beam all the way from amidships to the **transom**. Underwater hull shapes have also become increasingly shallow and both of these features combined tend to turn the yacht into a giant surfboard. Wide, flat hulls can also display a nasty propensity to slam into the waves when sailing hard to **windward**, which can make for a very wet and uncomfortable ride. Furthermore, a wide, flat stern can lift the yacht's rudder out of the water when heeled, causing it to lose its grip and allowing the yacht to 'round up' into the wind.

Less important, but almost equally frustrating, is the fact that a wide, flat stern allows little waves to slap annoyingly against the hull when anchored in any sort of a gentle chop or passing wash, which, as I know from my own experience, can often keep the occupants of the **aft** cabins awake at night, making for a very grumpy crew the next day.

Another design characteristic that significantly affects a yacht's speed through the water is the amount of hull below the waterline – its **wetted surface** to give it its correct term. The greater

Some modern cruising yachts have very beamy sterns

the surface area underwater – the greater the resistance to sliding through it. This can most easily be seen when comparing the deep-bodied hull of a long-keeler to a more modern fin-keeled yacht with a spade rudder and shallow hull sections. The underwater area of a long-keeler can be two or three times that of a racing yacht today, which is one of the prime reasons it is so much slower.

As with many things in life, the ideal solution for a non-racing boat is a compromise between the two and almost every cruising yacht now has a fin keel, spade rudder and extended beam aft. Creating a deeper vee to the forward section of the hull can tame, or even eliminate slamming to windward and slightly more rounded curves to the bilges aft can help avoid the worst of the slapping, as well as increasing her tracking stability (the ability of a yacht to keep to a course when sailing off the wind) considerably.

Keels

Knowing what type of sailing you want to do affects your choice of keel, which in turn helps narrow the range of yachts from which you can choose. Creek crawling and trailer-sailing is nigh on impossible with a racing yacht's deep keel. Likewise, **beating** off a **lee** shore with 15 degrees of **leeway** can be fairly heart-stopping.

Long-keeled yachts

A yacht is described as a long-keeler when the keel is an integral part of the hull. A long keel usually starts little more than a third of the way back from the yacht's **stem** and continues aft to join up with the rudder, effectively 'encapsulating' the propeller at the same time.

A hull design such as this would provide a comfortable ride in the choppiest of seaways. The fullness of the keel and hull tends to limit the amount of leeway (the tendency of a yacht to 'slide' sideways in the opposite direction to the wind) and also enables the boat to track downwind with little adjustment to the helm. Some swear by long-keeled boats for ocean cruising and, having covered many thousands of miles in a long-keeler myself, I would tend to agree – but only for long ocean passages and blue water cruising. As it is, now that I only get to pop out for a few days at a time from my base on the south coast of England, the ability of a more modern fin-keeled boat to manoeuvre in and out of tight, cross-tide marina berths is somewhat more important!

Fin keels

It really started when designers realised that cutting away the forefoot of a full-length keel took away some of the hull's resistance to tacking and turning, while only slightly reducing its ability to stay on track. This made the yacht less reluctant, and a good deal quicker to go through a **tack**. It also made her considerably more manoeuvrable in close-quarter situations under

This is typical of a long-keeled yacht made popular in the early 20th Century

power – particularly going astern. Furthermore, the reduction in wetted area lessened the hull's drag through the water, increasing its speed underway.

As usual, designers and naval architects experimented more and more with the reduction of keel length, realising that eventually they would reach the point where the keel's reduced **lateral resistance** compromised the yacht's ability to track in a straight line. Creating ever shorter keels meant that the effective ballast needed to be moved lower – i.e. the keel needed to be deeper – to create enough of a counterweight to the heeling effect of the wind on the sails. With a full-length keel the ballast was often only a relatively shallow strip of cast iron along the bottom of the keel itself. But reducing keel length whilst retaining the same amount of ballast meant it either had to be an all-iron keel, or the ballast had to be concentrated at the bottom of the keel, creating a bulge. The latter became quite popular as it had the added benefit of giving the yacht a more substantial base on which to take the ground, with the aid of 'legs' bolted to the topsides to stop her falling over.

Nowadays the same principle has evolved into the 'bulb-' or 'T-keel', whereby the majority of the yacht's ballast is concentrated into a torpedo-shaped bulb at the foot of the keel, while the rest of the keel is short and thin. This makes the boat very easy to manoeuvre both under power and sail, as there is so little lateral resistance offered by the keel, and it keeps the ballast where it is most effective – as low beneath the hull as possible. Whereas in the days of long-keel designs the ballast was often close to half the actual displacement of the yacht (an offshore yacht commonly had between a 40–50% ballast-to-displacement ratio), in a modern, bulbed fin-keel cruiser this can be reduced to around 30% because of the extra leverage and counterweight effect of having the ballast so deep – particularly if the ballast is made from lead.

Somewhere along the way designers also started playing around with the actual shape of the keel, trying to create the most hydrodynamically efficient sectional form that would create the least drag, but give the yacht a certain

Putting the bulk of the ballast low in the keel helps improve stiffness

amount of a 'lift' to windward. Clearly it isn't possible to use an aeroplane wing shape, as this is designed to only give lift in one direction. But it was soon discovered that certain shapes, namely making the keel slightly thicker forward than aft, reduced the turbulence created by the keel cutting through the water, and consequently its drag.

Winged keels

At some point in the late 1980s experiments with winged keels took place. Indeed, a few production boats, such as Sadler's Starlight yachts and Beneteau's First range, sported them for a while. The original idea behind the winged keel was to provide a shoal-draught boat with the ability of a deep fin to sail to windward by the use of hydrodynamics. Indeed, it was first played around with on the Americas Cup boats in the late 1970s and displayed a surprising ability to keep the yacht on track when sailing off the wind. On cruising yachts it could also create a small

The T-keel takes it to extremes, but the short keel improves manoeuvrability

dampening effect in a rough sea, slowing the up and down movement a little between waves. A winged-keel boat will also take the ground securely, but never without the aid of legs or props like those with twin keels can.

For production boats it was never truly successful or indeed popular and it rarely performed better overall than the plain deep fin keel – most likely because of the extra surface area and angle of attack, which inevitably increased drag.

Shoal keels

Shallow, or shoal-keeled yachts as they are more commonly called, became popular with those yachties who sailed regularly in shallow waters – the east coast of the UK for instance, or Chesapeake Bay on the eastern seaboard of the USA. Although there is a slightly detrimental effect on the

These keel fins are designed to act as efficient foils

yacht's ability to point (sail close to the wind) due to the lack of draught, the yacht's stiffness or stability under sail isn't unduly affected as the length (fore and aft) of the keel is extended in order to add slightly more ballast to compensate for lack of depth and ballast moment.

Twin keels

Twin or bilge-keeled yachts have several obvious advantages over those with a single fin keel, but also a few disadvantages as well. The first and most evident bonus is similar to that provided by the long keel – doubling the area that you can place the ballast means the keels can be made shallower. On a twin-keeler the amount of ballast required on each keel is roughly half that of a fin, resulting in two much shallower keels with each having to support less weight. This reduction in the yacht's draught is clearly a great benefit for coastal cruisers who like to sail in shallow

The shoal keel shown here keeps the draft to a minimum without loss of stability

waters. Furthermore, a twin-keeled yacht will take the ground upright and securely without the need for supporting legs. So if drying out on a beach is your particular fancy or simply sitting on the hardstanding without the need for a cradle, then you might want to investigate twin-keelers a little further.

There is, however, a drawback to yachts with shallow twin keels. The amount a hull will slip sideways in the water (leeway) is proportionate to the area of the keel being presented to the water. The bigger this area, the less it will be inclined to leeway. Now, you might think that two keels means a greater area of keel to provide this resistance – but it doesn't work like that. Only the one keel is actually providing resistance to leeway under sail – the **leeward** one – so it soon becomes clear that having two much smaller keels instead of one large one will allow the angle of leeway to increase – often in the region of an extra 5–10 degrees.

On early bilge-keelers, where the keels are angled considerably away from the vertical (i.e. splayed out looking from the front or rear), there is a tendency, when the boat is well heeled over, for the windward keel to lift clear or nearly clear of the water. This causes the keel to slam noisily and uncomfortably into the oncoming waves as well as increasing the sideways loading on the keels. This was often the cause of leaks at the keel-hull joint in early models and it was common to see stress cracks running out from the keel area of the hull. Later models had the keel-root area reinforced – often with the addition of reinforced, solid GRP keel stubs.

After a while producing boats with two keels, some designers (most notably David Thomas of British Hunter Yachts fame and David Sadler of Sadler Yachts) realised that with two keels you could actually use the shape and angle of the keel to go some way to counteracting the extra lee-way caused by the lack of draught. Their twin-keel yachts had 'foiled' or 'asymmetric' keels that were positioned on the hull so as to point a few degrees in towards the hull's centreline. They

Boats with bilge keels can be put on a trailer and will take the ground easily

were also more upright and slightly closer together than those on previous bilge-keeled boats, which to some degree got over the slamming problem of the more splayed bilge keels. The twin-foiled keels also had the effect of pushing the hull up to windward when **reaching**, which duly countered some of the loss of tracking angle caused by the yacht's leeway.

Of course other designers soon jumped on the bandwagon and the expression bilge-keeled soon became the more respected twin-keeled, indicating that the latter had foiled keels, which were superior to the old dual keels that were just bolted onto the bilge for reduced draught and increased convenience.

Retracting keels

Most often found on dinghies and trailer-sailers (on a dinghy it is called a dagger-board), retractable, or 'lifting' keels as they are commonly called, offer numerous benefits to the cruising yachtsman.

There are two main types of retractable keel – lifting and swinging. Lifting keels are simply a larger and heavier version of the daggerboard on a dinghy and are usually raised by some form of mechanical winch and strop into a sealed keel box, which often impinges on the accommodation below on smaller boats. On older boats you were generally required to jump below and start winding madly to raise it, but on more modern designs the lifting tackle is led up through the deck and onto a halyard winch or similar. This has the added advantage of allowing a single-hander to lift the keel as he is approaching shallow water.

A swinging keel, or **centreboard** as it is sometimes known, is hinged at the leading edge and you raise it via a **line** attached somewhere along the trailing edge of the board. The advantage this has over the lifting, daggerboard-type keel is that the centreboard can

Some lifting keels can really impinge on the boat's accommodation

bounce up harmlessly should you hit the bottom unexpectedly, whereas the lifting keel will plough its way through the seabed just as a fixed keel would, until it is winched up. Worse than grounding with a fixed keel, hitting the bottom with a daggerboard-type keel can cause damage to the lifting mechanism or keel box itself.

Some swing-keel yachts, or centreboarders, have a shallow-draught, ballast keel 'stub' into which the swing keel retracts. This means the draught can still be reduced considerably, but with no keel box or similar intrusion into the accommodation. Obviously a yacht with this arrangement will draw more with its board lifted than those with keels that retract entirely into the cabin, but it's often a useful compromise, particularly when loading the yacht onto a trailer.

Some retractable keels or centreboards are 'ballasted' in that they are made from heavy iron or steel. Others are just a fairly lightweight board designed to reduce leeway rather than having any ballasting effect.

The former can be useful in a small cruiser for a couple of reasons. Firstly, when they are fully extended down they add to the stiffness of the boat under sail. Secondly, a heavy retracting keel is much easier to lower if it has some weight in it and the force of gravity is on your side. This can be particularly relevant if the boat is left on a drying mooring and takes the ground between tides, as bits of gravel, sand and shell can often find their way in between the plate and the keel box after a while. Barnacles and other growth will also appear if the board isn't properly painted with antifouling.

There are some larger cruising yachts that have retractable keels, most notably the British-built Southerly yachts and the French-made Ovni and Feeling. Southerlys have always been fitted with swinging keels, only nowadays they are raised and lowered electrically, rather than by hand. The keel box is actually a huge ballast plate, which is securely bolted into a precisely shaped recess in the bottom of the hull. The keel itself is weighted a little, but rarely makes up more than a small fraction of the yacht's overall ballast.

Other lifting-keel yachts, such as the Feeling range, have cast-iron ballast bonded into the inside of the hull using GRP. The great advantage of Southerly's ballast plate, however, is that it is designed to take the knocks and scrapes of grounding without the risk of damaging the GRP hull.

The Ovni range of aluminium boats built by Alubat on the French Atlantic coast, all have an electro-hydraulic, push-button lifting keel and a single hydraulic rudder that is raised using a hand pump in a cockpit locker. Rudders can be a problem on retractable-keel yachts in that they clearly need to be short enough not to be damaged when the boat takes the ground, but long enough to do their job and keep a grip on the water when the boat is heeled over. The popular design these days is to use twin rudders, which not only solves the draught problem, but also affords other benefits such as ensuring the rudder always remains in the water and as vertical as possible when heeled, to enhance the sensitivity of the steering.

The Southerly lifting keel showing the ballast set into the hull moulding

As with swing keels, rudders on these types of yacht can also be of the swinging type, but this is more usually found on smaller boats where the rudder is hung on the transom and can be retracted simply by tugging on a line.

Water Ballast

Occasionally you'll come across boats that use seawater as ballast. This is nearly always for trailer-sailers rather than cruising yachts, unless it is being used in addition to regular fixed ballast, to trim the boat.

In the last 10 years or so there have been quite a few 'power-sailers' introduced to the market place – MacGregor, Hunter/Legend Edge and Odin/Imexus to name a few. These use a water-ballast tank that is filled with seawater after launching, usually by opening a simple **seacock** or valve. While this might seem an eminently sensible idea, in that the ballast can simply drain away on the slipway before you drive off, water ballast is by no means as effective as high-density ballast such as iron or lead. For a start water weighs nothing in water, so the only counterweight effect is when it is lifted up above sea level. Imagine a plastic jerrycan half-filled with seawater. If you held it upright in the sea it would float so that the water levels inside and out are equal. Now, if you try to 'heel' the jerry can, you will feel absolutely no resistance – in fact if you let go completely it will simply fall on its side.

What this shows is that the water as ballast does nothing unless it is contained in a tank, which itself is lifted above sea level when the boat heels. The lower the tank is in the boat the less use it is, as only the portion of the tank that is lifted clear of the water level outside, probably around a third, will be acting to right the boat. But to self-fill, as they are all designed to do, the tank must be below sea level. This means that at low angles of heel the boat will be fairly tender (prone

The Imexus power-sailer can reach speeds of 20 knots without its water ballast

to heel), not stiffening up until the hull is heeled sufficiently and has lifted enough of the water ballast tank above sea level to create a counterbalance effect.

If you were to pump the water into two tanks, say, mounted immediately above the waterline on each side, you would get the maximum righting moment immediately, as the entire quantity of water ballast in the tanks becomes effective as soon as the hull starts to heel by as little as just 1 degree.

It's also worth bearing in mind that, since 1 litre of seawater weighs approximately 1 kg, and that an average 24-footer, say, usually has around 500 kg (half a tonne) of iron ballast, the size of tank required to provide the equivalent amount of water ballast would be 500 litres (half a cubic metre) – or enough to fill a single tank measuring 4 m long × 1 m wide × 0.25 m high. Despite this being pretty impractical for such a small boat, MacGregor Yachts somehow manages to squeeze 550 kg of water in its 26 ft power-sailer (plus 130 kg of iron) and the Odin 820 (27 ft) a mammoth 800 litres along with the same 130 kg of iron ballast!

It's fair to say that water ballast alone is not sufficient to keep a sailing yacht stiff, stable and upright in a seaway. Most also carry some form of high-density permanent ballast such as iron, which supplements the water ballast – a sensible compromise to my mind as it helps provide the necessary initial stiffness that the water ballast can't until it is lifted clear of the sea level.

Since their inception, and after a few rather unfortunate incidents when the owner has forgotten to fill the water tanks before hoisting the sails, most of these water-ballast tanks are designed to fill automatically on launching, requiring the valve to be consciously shut to stop it from taking on ballast.

Rudders

Much as keels have developed enormously over the past century, so too has the design of the humble rudder. Originally, rudders were attached to a hull using **gudgeons** and **pintles** (hinges basically) – usually one at the top on the transom and one at the bottom on the back end of the keel. Larger rudders had more pintles where required, but the principle remained the same. While this worked successfully for many years, the biggest problem with the transom-hung rudder, as it is known, was that the steering was very heavy. All the force of the water passing over it had to be diverted in order to make the yacht's stern turn a few degrees, which required a fair bit of elbow power!

At the time the fin keel was being developed designers realised that, in stopping the keel short of the rudder aft, it was going to be necessary to provide additional support for the rudder, now that it could no longer hang off the back end of the keel. Hence, a sort of mini-keel was created, called the skeg, which hung down from the stern and provided a rigid support for the rudder pintles. At this point yachts of this design were commonly referred to as 'fin and skeg' hulls.

While this might have solved the dilemma of where to hang the rudder, the problem of heavy steering still remained – until, that is, some bright spark realised that if even a small part of the rudder's surface were placed forward of the hinge line (i.e. the skeg was shortened slightly to allow some of the rudder to overlap below it), then as soon as the rudder was moved off the hull's centreline, some of the force of the passing water would be pushing against the small overlapping area of rudder and actually assisting the helmsman in steering – a sort of primitive power steering. This effect became known as helm balance and was almost immediately adopted by all the major production boatbuilders.

Designers continued plugging away at this subject and it wasn't too long before they realised that they could do away with the skeg entirely by building

This half-skeg provides a strong rudder mounting and lighter steering

the rudder around a very strong stock, which was supported top and bottom by two big bearings built into the boat. Thus the spade rudder was invented, which greatly improved the precision and reaction time of the helm, as well as taking much of the weight off. Now, all the water immediately flowed over the rudder, rather than coming off a skeg, and the balancing portion could be made along the entire leading edge of the rudder. This made the yacht very light to steer, but offered better feedback to the helmsman than ever before.

Many were and still are concerned about the lack of a skeg to offer support to the rudder. Indeed there are times – when in collision with a semi-submerged object for instance – that the skeg can protect the rudder from damage, or at least stop floating lines getting caught between the top of the rudder and the hull. However, the spade rudder has today been adopted almost universally on production yachts, with only a handful of blue water cruising yachts retaining a skeg. But it might be worth taking into consideration if you're planning some serious ocean cruising.

Along with dual helming stations, another popular development on modern yachts is the use of twin rudders. In addition to the reduction in draught – a particularly useful feature with lift keel-yachts as mentioned earlier – twin rudders are a notable benefit to modern cruising yachts, which tend to have very beamy sterns. A wide stern with a sharp turn of the bilge might be fine for downwind surfing, and for cramming in a second double aftercabin, but when heeled over going to windward, a regular single spade rudder tends to be lifted out of the water, causing the yacht to lose its grip and spin on its keel towards the wind – rounding up as it is commonly called. With twin rudders, however, it doesn't matter if the windward one lifts completely out of the water because the one doing all the work, the leeward rudder, will now be deep down and – better still – pretty much vertical, making it far more effective and causing less drag on the yacht.

A typical bow thruster helps manoeuvre her bows into tight berths

Where twin rudders become a problem is when manoeuvring the yacht under power. A propeller positioned centrally just forward of the single, central rudder, forces its wash over the rudder's surface, enabling you to change direction quickly and easily with a short, sharp blast of the throttle.

With twin rudders – one at each side – the prop wash in the centre doesn't go over the rudders at all, but right between them. So it's not until the boat is actually under way with water flowing over the rudders that you have any steerage. The way boatbuilders have overcome this problem is by fitting a bow thruster – a small (usually electric) motor is installed transversely through a short tunnel in the bows of the yacht that provides side thrust in either direction to help steer the yacht at low speeds under power.

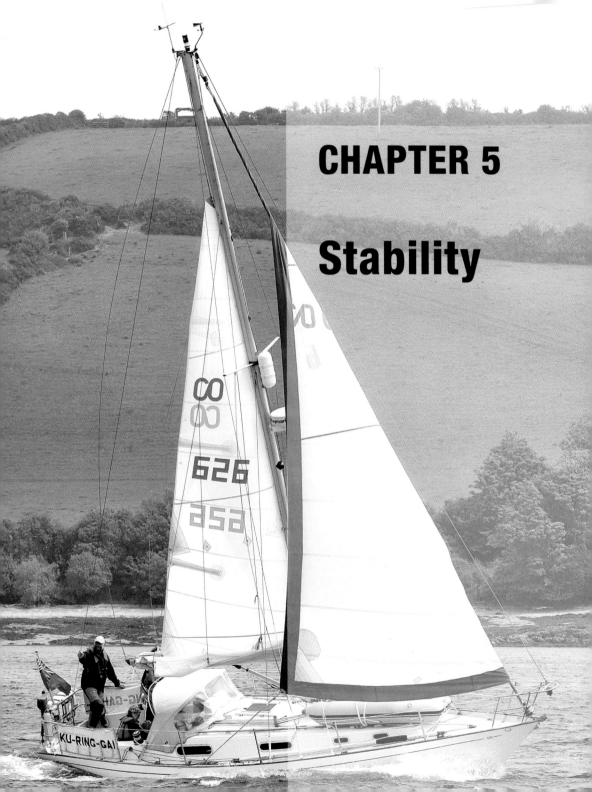

CHAPTER 5

Stability

European Recreational Craft Directive (RCD)

In 1998, the EU trade authorities introduced Europe-wide standards for the boatbuilding industry to uphold when designing and building new sailing yachts and motorboats. The Recreational Craft Directive (RCD), as it is known, applies rigorous minimum standards and specifications to all leisure craft, in order to ensure that they are safely capable of withstanding certain wind and sea conditions.

Boats built to comply with the RCD (self-built craft don't have to providing they're not sold for five years after launching) are CE-marked and carry an identification mark known as the Craft Identification Number (CIN). Similar to a car's VIN plate, a builder's plate must then be supplied on every vessel launched, which clearly displays the CE mark and the category(ies) for which it was designed.

In fact, usually only one sample prototype actually goes through the testing process, after which it is assumed all the production yachts of this particular model will be built to the same standard as that tested. So the onus is then on the boatbuilder to ensure it continues to build that model to the exact specifications that were passed during the RCD trials.

After testing, the boats are classed within a particular category, of which there are four:

A – Ocean: Designed for extended voyages where conditions may exceed Force 8 (Beaufort scale) and significant wave heights of 4 m and above, but excluding abnormal conditions. The vessel is largely self-sufficient.
B – Offshore: Designed for offshore voyages where conditions up to, and including, winds of Force 8, and significant wave heights up to, and including, 4 m may be experienced.
C – Inshore: Designed for voyages in coastal waters, large bays, estuaries, lakes and rivers where conditions up to, and including, winds of Force 6, and significant wave heights up to, and including, 2 m may be experienced.
D – Sheltered Waters: Designed for voyages on sheltered coastal waters, small bays, small lakes, rivers and canals when conditions up to, and including, winds of Force 4 and significant wave heights up to, and including, 0.3 m may be experienced, with occasional waves of 0.5 m maximum height, for example wash from passing vessels.

'That's a good idea isn't it?' I hear you say. Well, in some ways yes – in that it lays down a few basic rules about stability that boatbuilders must adhere to when designing a yacht for certain conditions. But the worry is that those new to the sport of sailing might take them as gospel – i.e. this is a Class A boat, so I can sail it in all weathers and sea conditions. Sadly, in almost all cases where a regulated standard is introduced into any manufacturing business, the maker is tempted to 'work down' to the lowest acceptable specification in order to save money.

For instance, the stability element of the RCD – the Stability Index (STIX) – is measured by mechanically heeling the boat over in a test tank to determine the level of self-righting ability the boat has

A typical builder's CE plate clearly showing the RCD category of the vessel

at a certain angle of heel, right up to the point at which the boat overturns, known as the Angle of Vanishing Stability (AVS). However, this is done 'light' under laboratory conditions without the sort of loads that any normal cruising yacht might have aboard on a long passage. So the water and fuel tanks are empty, there's no crew on board, no heavy gas cylinders, no gear up the mast such as a radar scanner, reflector etc. and no dinghy, outboard, barbeque and all the rest of the usual paraphernalia that cruising yachties take along with them. In 90% of cases, were you to carry out a STIX test on a fully-loaded cruising boat, the results would differ significantly to those found by testing an empty boat – possibly (quite likely in many cases) to the point where it would no longer be rated within the category in which it is licensed to operate. Just to emphasise my point, new yachts must have a plate to show which category the boat has attained, which frequently states something like A-6, B-8, C-10. These figures mean that this yacht is a Category A craft, but only when six people are on board. Put two more on and it becomes a Category B vessel, four more and it's a C-rated boat.

A yacht undergoing heeling tests to ascertain her self-righting ability

One of my jobs for the yachting media involves test sailing new yachts and writing reviews about them and my findings. I used to include the STIX figure of each boat tested in the specifications table, until I realised just how misleading and meaningless they actually were. Instead, when I'm test sailing the yachts I deliberately over-canvas them in strong winds to see how they cope in a real-life situation, rather than believe a set of figures obtained under somewhat questionable conditions.

I'm not saying that any of the boatbuilders are breaking the law, or intentionally endangering the lives of their clients, but it is quite understandable in any regulatory system that some cost-conscious manufacturers will take the view that the barest minimum will do. For this reason, a new lightweight production yacht often rates at just 0.1 over the STIX level required for the yacht to be within a particular category. Many Category A-rated (Ocean) yachts, for which the STIX must be 32.0 or above, come in at 32.1–32.2. So you've only got to put a 2 kg radar dome halfway up the mast and that STIX figure will probably drop to 31.8–31.9 – in theory demoting it to a Category B vessel. Food for thought is it not?

Ballast Versus Form Stability

In a sailing vessel there are two principal methods of preventing a yacht from heeling too far under sail, or capsizing totally. The first is by incorporating a heavy weight, known as ballast, into the bottom of the vessel to counteract the force of the wind acting on the sail. This ballast will typically be in the form of a high-density material such as iron, steel or lead and is usually attached to the yacht in the form of a keel. Alternatively, it can simply be fixed or bonded into the bilges of the boat – commonly found in lifting-keel yachts. The quantity of ballast needs to reflect the amount of sail that could be hoisted and the exact position of the ballast will depend on the height of the rig and where the wind's **centre of effort** is on the sails. Clearly, a large ballast weight attached to the bottom of a keel that extends several metres below the boat will withstand considerable force on the sails and keep it from heeling over uncomfortably far. Whereas ballast placed inside the hull – in the bilges for instance – will have far less righting moment (leverage required to counter the force of the wind in the sails) than that fixed at the foot of a deep keel. So in the case of inboard ballast, either the quantity of ballast must be duly increased, or the height and size of the sails and rig must be reduced accordingly.

One of the benefits of a lifting keel – but at the expense of righting ability?

Boats with lifting keels have long suffered criticism about their 'tenderness' (tendency to heel excessively in strong winds). The reason for this is that these boats have to carry the bulk of their ballast in the hull, rather than in their keel. Consequently either more ballast than usual is required in the form of a large iron base-plate, iron stub keel or even just big bars of iron fixed into the bilges – or the sail area/rig height must be reduced to compensate. Some lifting-keel yachts have ballasted centreboards made from iron or steel, but there is a limit as to how heavy they can be before the job of lifting them becomes impractical.

The second method of inducing resistance to heeling in a sailing yacht is by creating a hull shape that is naturally more stable in the first place – ballast excluded. Now this is where it gets complicated and the yacht designer really starts earning his salary. A 20 ft X 20 ft, 0.5 in thick plank of wood placed on the surface of the water would have enough inherent stability to enable it

A wide, flat yacht like this has masses of form stability to keep her upright

to support a smallish mast and sail without the need for any ballast. This is because it naturally wants to stay flat on the water, providing some – albeit limited – resistance to being turned over.

Bearing in mind this rather extreme example of the principle of what is commonly known as 'form stability', it then becomes fairly obvious that if a yacht were made very wide and flat it would display more of an innate tendency to remain upright than if it were tall with a narrow hull.

If the ability to stand up to its sail were the only design factor for a yacht, the job would be easy. But clearly a yacht needs to be able to use the power of the wind as efficiently as possible to drive it forward, so a plank with a stick is simply not going to work!

The designer's job is to create the ideal compromise between stiffness and performance, which is an extremely complex task involving myriad calculations and compromises along the way. I'm sure yachts would look totally different if, like the flying drones the Forces use for surveillance in the battlefield, they were unmanned. Without a crew eating, sleeping and generally living on board, clearly a far more stable yacht design could be created. But that isn't the case – a sailing yacht will always be a compromise and no matter how good the design is it will always be biased towards one function or another to a varying degree.

What you need to decide before buying a boat is in which direction you would prefer this bias to be. Do you want a yacht that is incredibly stiff, but slow and lethargic in the water? Or would you prefer something a lot twitchier that planes off the tops of waves at twice its displacement hull-speed? Obviously I'm putting forward the extremes here for the purpose of illustration, but you will need to have a fair idea of your own preferred balance between stability and performance before you waste hours traipsing around the boatyards looking at boats that simply won't ever fulfil your wishes.

Vanishing Stability

I mentioned earlier the 'Angle of Vanishing Stability' (AVS). If you've been reading magazine or Internet reviews on potentially suitable yachts you may have come across this expression. Indeed some yacht reviews publish a little graph along with the specifications showing what is commonly called a 'GZ curve'.

What this curve or graph shows is the righting moment (the force acting against that which is trying to capsize it) of a yacht as it is gradually heeled over, degree by degree, through a complete 360-degree rotation. It is designed to accurately represent the self-righting capabilities of the yacht in the event of a total knockdown, or a '360' as it is often called.

The first upward trend of the GZ shows the self-righting force of the ballast and form kicking in very quickly at smallish angles of heel – i.e. usually between 0–45 degrees. Then as the heeling

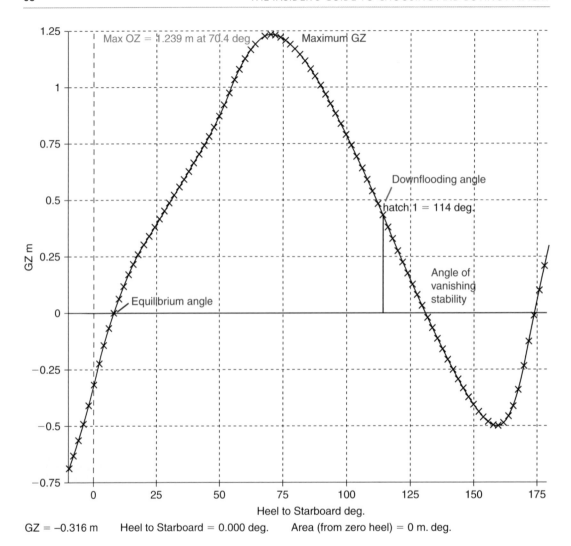

GZ = –0.316 m   Heel to Starboard = 0.000 deg.   Area (from zero heel) = 0 m. deg.

A typical GZ curve of a sailing yacht

angle increases still further it soon reaches the point of maximum righting moment (around 60 degrees is common). After this point the curve starts to fall away back down towards the zero GZ line, showing that the effect of the ballast and hull shape is now falling away – i.e. the forces on the rig and hull from the wind and waves are slowly defeating the yacht's self-righting ability. The steeper this curve is, the quicker the yacht is losing its stability. Ideally it should be a smooth curve much like the first, rising part of the graph, but sometimes – often in very extreme racing boats such as Open 50s that are very wide and flat – this can drop away almost vertically. This indicates that once the yacht is past its point of maximum righting ability it is almost guaranteed to flip over very shortly after. Saying that, the fact that today's racing boats are so flat and shallow-hulled means the initial rise of their curve will also be very steep, as these designs are very resistant to being heeled at first, but quickly flip over once they do reach the critical maximum GZ point.

The point at which the falling curve cuts through the zero GZ line indicates the AVS – in other words the point at which the yacht has run out of positive righting ability and will now capsize completely. This, along with the angle of the curve above the zero GZ line, are good indicators of just how resistant a yacht is to turning turtle. A lightweight production yacht with a shallow hull and average ballast will probably reach its AVS (likely to be somewhere around 120–125 degrees of heel) around 10–15 degrees before a heavier, deep-vee hulled, fin-keeled blue water cruising yacht, and will usually display a steeper falling curve.

However, that's only half the story. Although the curve above the zero GZ line shows the yacht's resistance to being turned over, the curve below the line shows the yacht's ability to right herself

Vessels with large, buoyant superstructures are highly disinclined to turn over

after a total capsize – an even more critical indicator of a yacht's suitability for open-ocean sailing. What you really want to see here is as steep a curve as possible, but one that only just dips below the line, meaning that the yacht is inherently unstable upside down and considerable forces are acting upon it in order to return it to the mast-up/keel-down attitude.

It is important to realise that the self-righting forces on a yacht are not entirely the result of the weight of ballast in its keel, but in the shape of the entire yacht – both above and below the waterline. A shallow, flat hull with a sharp turn at the bilges will possess a great deal of form stability and will resist heeling to a considerable degree at first. However, once it has gone through its AVS and flipped over it can almost be as stable upside-down as it was the right way up because of the lack of **buoyancy** in its hull above the waterline.

A boat with a tall, chunky pilothouse on deck, however, will display a greater propensity to turning back upright because of the inherent buoyancy provided by the air within this superstructure. It will usually be much quicker in righting itself than a flush-decked yacht – provided the deckhouse doors or washboards are closed that is!

A typical racing multihull that is regularly pushed to the limits

A more conservatively proportioned cruising multihull

Multihull Versus Monohull

The example of the 20 ft X 20 ft wide plank being highly stable is even more relevant when study-ing the design of multihulls and their resistance to capsize. A multihull doesn't rely on ballast to keep it upright, but on the buoyancy of the leeward hull (or float in the case of a trimaran) and the distance it is from the centre of the wind force – i.e. the foot of the mast. The wider the beam the more resistant it will be to turning over. Eventually what can happen under very extreme cir-cumstances is that the windward hull will lift out of the water, and if this force isn't very quickly reduced by slackening off the sheets or turning away from the wind, there's a good chance she will capsize, and quite quickly.

Of course, once a large multihull has capsized it's virtually impossible to right her at sea as the forces required to do so are fairly massive. The good news, however, is that this is very unlikely to happen in any modern beamy multihull and, even if for some extreme reason it did, these types of craft don't sink. Thanks to the lack of ballast and generous buoyancy they just continue to float – only upside-down.

There have been rumours doing the rounds for many years regarding the stability of multihulls and whether they are particularly prone to being overturned in rough seas. These rumours have been substantiated somewhat by a few high-profile cases of racing multis flipping, or more likely trip-ping over themselves, in ocean storm conditions. But these instances are extremely rare nowadays and most reported incidents are caused by racing crews pushing their vessel to the absolute limit of its capabilities. Capsizing is almost unheard of with cruising multihulls these days – fortunately.

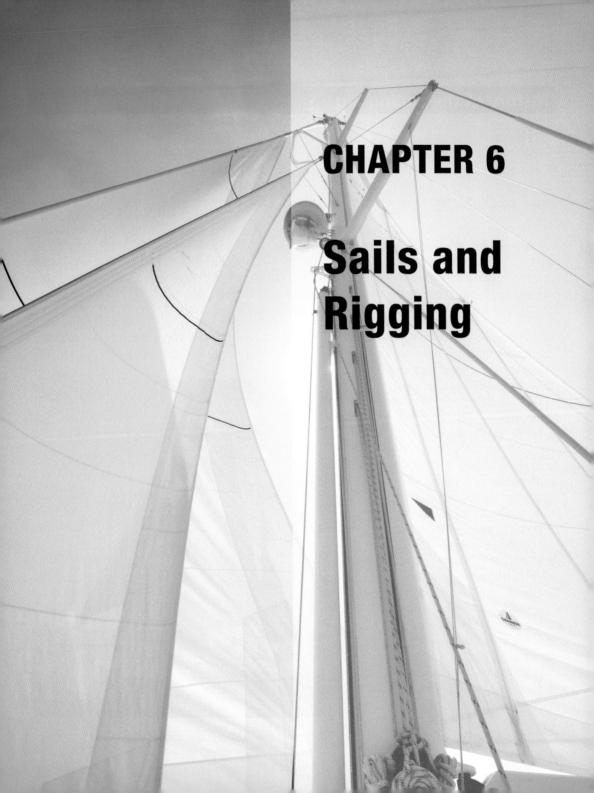

CHAPTER 6

Sails and Rigging

History of the Sailing Rig

To understand how today's modern rig came about it is necessary to look back at the development of the fore-and-aft rig over the last few centuries. Early sailors chose a square sail because it was easy to make and simple to control on a centrally-positioned mast. Whilst essentially a downwind sail, it was modified gradually over the years to become more efficient and allow it to make some sort of headway other than dead downwind.

The basic square sail was modified to become the windward working lateen rig, which could take a vessel to windward of a sort, but was a handful to control on larger boats. By hauling the yardarm behind the mast, the lateen became the first effective fore-and-aft sail plan for boats. But the long, dipping yard meant that either the sail had to push against the mast when changing tack, or the yard had to be 'dipped' (tipped almost vertically in order to feed the yard around the back of the mast) to allow the sail to remain on the leeward side of the mast.

In an attempt to overcome the problems with this unwieldy sail, various additional modifications were made to the single sail, or una rig, resulting in the lug, cat and **gaff** rigs. The lug-sail utilises a top yard that only just sets forward of the mast, allowing the yard and sail to be dipped and tacked much more readily. This modification soon saw the appearance of the cat and gaff rigs, wherein the yard and **boom** were actually attached to the mast using rope hoops or wooden jaws. Cat and gaff rigs are similar in that they feature a single, quadrilateral sail with **luff** and leech, supported between an upper yard (gaff) and a boom. In both rigs the mast is placed well forward in the hull in order to balance the forces on the helm and this presents problems with staying the mast itself. If the distance between the mast and the bows is shortened, the stays will be attached to a narrow part of the hull. This reduction in **shroud**-base width lessens the effectiveness of the mast support and overall rig stability. For this reason, boats with this type of rig tend to have very broad shoulders and

A fine old gaff-rigged cutter under full sail

a plumb stem to allow increased distance between the deck and the mast heel (usually in the bilge).

The single-sail rig is really only controllable on hulls up to 36ft, above which the sail becomes almost unmanageable. It was soon realised that moving the mast back towards the centre of the hull and attaching a sail to the **forestay** would retain the balance and driving force, but reduce the size of the individual sails, making them easier to handle. Thus the fore-and-aft rig was created in its varying forms. At the same time it was realised that splitting the sail forward of the mast into two or more sections on larger boats would further reduce the headsail size and allow better control of the centre of effort on the rig in heavy weather. Hence the cutter rig came about, wherein a second headstay – the inner forestay – carried a much smaller sail called the staysail. The foremost sail, the **jib** (later to become known as the yankee), was cut with a high **clew** to facilitate better visibility forward and to allow greater airflow to the staysail. It also created a **slot** effect between the two sails, which forced the air passing between them to accelerate.

A modern cutter rig gives this Island Packet plenty of options

As yachts grew larger, boatbuilders adopted the multi-mast approach, to reduce the area of each sail still further. Hence, the introduction of the ketch, yawl and schooner rigs.

Development of the Modern Rig

The now ubiquitous, high-aspect (tall and narrow) Bermuda rig came into being at around the turn of the 20th Century, after the fashion for leg-o'-mutton **mainsails** (one in which the top yard was virtually vertical so as to form an almost triangular sail) had created a preference for triangular mainsails. It was soon realised that if the luff of the mainsail was attached directly to the back of the mast, the extra weight of the **spar** aloft could be eliminated, further aiding sail handling for smaller crews and allowing for taller masts. Also, the gaff rig jaws meant the top yard (gaff boom) could not pass beyond the spreaders, thus forcing the upper part of the mainsail (topsail) to be loose-luffed. Whereas, with a sail track reaching to the masthead, a full-height mainsail could be hoisted in which the entire luff could be firmly attached throughout its length. Although this sail plan raises the centre of effort and reduces stability initially, the triangular sail has the benefit of being more easily controlled. Reducing sail area by **reefing** down can be done rapidly and lowers the centre of wind effort very quickly due to its shape. For these reasons, the **Bermudan** (aka Bermudian) sloop rig as it became known, has become almost universal in today's small to medium-size cruising yachts.

The Bermudan-rigged sloop is now the most popular sail plan

Of course some traditional vessels still sport attractive gaff rigs, and even several new ones, but these are mainly for the enthusiast and rarely the chosen sail plan for today's leisure sailor.

Junk rig

One rather more unusual rig that has been around for centuries and is still in use to this day is the **unstayed** Chinese junk rig – although usually in the form of the more modern version pioneered by the famous sailor and adventurer, Blondie Hasler. Said to be an ideal rig for single-handing, Hasler designed it specifically for his highly modified Folkboat, *Jester*, in which he crossed the

Atlantic on the first Observer Single Handed Transatlantic Race (OSTAR) in 1960. The junk-rig sail comprises a single sail with a yard at the top, a boom at the bottom and full-width **battens** every couple of feet. It is attached to an unstayed, flexible mast using simple parrels on each batten. The single-sail version is commonly known as the cat rig, although numerous ketches, yawls and schooners have also been built using junk-rigged sails.

Junk-rig enthusiasts say that the design offers a more comfortable ride as the flexible mast gives a little in the gusts, which prevents excessive heeling. It is also simple and safe to handle, given that all the sail controls are led into the cockpit. Tacking simply involves putting the helm over – the sheet can be ignored – and reefing just requires the halyard to be let off and the weight of the top yard and battens will ensure the sail folds neatly down onto the boom like a Venetian blind. Wear and

A junk-rigged yacht is easy to handle and simple to reef

tear is low in the junk rig as everything seems to be under calm control at all times, rather than the noisy and somewhat scary headsail-sheet antics performed each time a Bermudan sloop-rigged yacht goes through a tack or **gybe**.

One of the most significant benefits of the junk rig is its simplicity. There is no **standing rigging** to be concerned about or maintain and far fewer expensive turning **blocks** on the **running rigging** to keep an eye on.

The downside of the junk rig has always been its poor performance to windward, but recent developments with sail cut and batten design have gone a considerable way to overcoming this problem.

Unusual rigs

One or two other interesting rig designs have also appeared over the past 40–50 years, but have not proved particularly popular for one reason or another. Garry Hoyt's Freedom rig is a typical example, where the sail is wrapped around an unstayed mast and held flat using a wishbone boom.

Another unusual rig that never really took off was Carbospars' Aerorig, in which the sails were fixed to a rotating mast with a fixed boom that stretched forward and aft. Almost none of the more adventurous and innovative rigs have been adopted by mainstream production boatbuilders.

Of course high-tech rigs and sail designs are always being experimented with amongst the racing fraternity, but the costs are usually too prohibitive to allow many of these ideas to trickle down to the average leisure-boat owner. The majority of developments taking place over the last few decades have been centred on ease of handling. The furling headsail has revolutionised sail handling for the short-handed crew and similar modifications to mainsail reefing are being worked on to the present day. This, combined with much improved deck gear, has made going to sea in a 40-footer with just two on board a perfectly viable and safe option.

The Freedom rig had unstayed masts of aluminium or carbon-fibre

Masthead Versus Fractional

During the 1960s and 1970s certain peculiarities came about, thanks to the International Offshore Racing (IOR) rules, which tended to distort both rig and hull into a 'diamond' shape. The complex IOR rules evolved from the Cruising Club of America's (CCA) rule for racer/cruisers and concentrated on hull shape, length, beam, freeboard and girth dimensions, as well as **foretriangle**, mast and boom measurements, and inclination stability. The measurements were used to compute the handicap, which is an 'IOR length' in feet. In a handicapped race, the IOR length is used to compute a time allowance in seconds per **nautical mile**, which is multiplied by the distance of the race, and subtracted at the race end from the boat's actual course time, to compute its corrected time. The longer the IOR length, the smaller the time allowance. Mainsail area attracted less IOR length than jib area, which led racing boats to adopt the fractional rig and hulls with pinched ends, wide beams, excessive tumblehome and retroussé sterns; these changes did little to improve the boat's handling characteristics, but were merely an effort to 'cook' the books and give the yacht a favourable rating. The IOR system came to grief after the disastrous Fastnet Race in 1979, when many racing yachts and their crew were lost in extreme weather conditions.

The increasingly popular fractional sloop rig, which has the forestay terminating some way down from the top of the mast, has now become *de rigueur* for small to medium-sized cruising sailboats. Usually expressed as a fraction, i.e. nine-tenths, or seven-eighths, depending on how high up the head of the stay is mounted on the mast, this development was first tried and tested by the racing fraternity, as it gives more

The stern of a typical IOR-designed boat of the 1970s

control in the bending or raking of the mast and hence more flexibility with adjustment in mainsail shape under way. The masts are also placed further forward in the hull, which means smaller, more easily handled jibs can be used.

The masthead rig, still common on some offshore and bluewater yachts, is simpler and less fussy about angles, mast bend, **rake** etc. It could loosely be termed as a 'fixed' rig, wherein the angles are mostly preset and not likely to be changed at sea. The fractional rig, however, is far more tweakable and usually has mast-bend control through an adjustable **backstay** to change the sail shape and adjust to the conditions and wind angle.

Masthead rigs usually generate their main source of power from large (up to 150 %) overlapping **genoas** and consequently have smaller, easier-to-control mainsails. This is ideal for short-handed

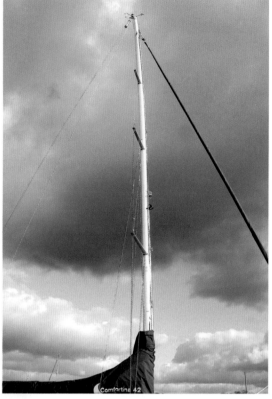

The masthead rig was commonplace for cruising yachts at one time

The forestay of a fractionally rigged yacht is some way below the masthead

blue water sailors who often need to reef alone at an ungodly hour. The boat is easily de-powered by simply furling the genoa a few turns, then if conditions continue to worsen, the mainsail is small enough to reef without drama – especially if it is fitted with single-line reefing back to the cockpit and lazyjacks to catch the sail neatly before it falls over the deck.

The proliferation of the fractional rig with its large mainsail has come about partly by the desire for better performance, but also the wish for smaller, more easily controlled headsails. Nowadays designers tend to maximise mainsail area in order to increase downwind and light-airs performance, so it is important to be able to induce some bend in the mast, pulling the top section aft by tightening an adjuster attached to the backstay. This increases overall stability when going to windward in brisk conditions by flattening the mainsail, moving the draught further forward and opening the

leech. Off the wind the backstay is slackened to put more camber back into the sail and in light airs the large main offers plenty of controllable power.

The rig can be tuned overall to the conditions you generally sail in. Designer and racing yachtsman, Dudley Dix, advises 'for more power from the mainsail in light winds slacken the caps a bit and tighten the lowers (and intermediates if a multi-**spreader** rig) so the mast will stand straighter. If regularly sailing in strong winds, pre-bend the mast some more by tightening the caps and slackening the lowers and intermediates. This will pull the luff of the main forward and flatten the sail to reduce power.'

Although the straightforward masthead rig is generally thought of as more suitable for offshore cruising – especially with a short-handed crew – the ability to tweak a fractional rig for optimum speed and efficiency has become the modern trend, allowing the boat to be used for racing as well as comfortable family cruising.

A further advantage of the fractional Bermudan rig with its smaller headsail is the ease with which it can be tacked. With no forward lower shrouds, baby stay or inner forestay to get in the way, and no large overlap, the sail can be whipped across the boat in very little time and with the minimum of winch grinding. Obviously, for racing this is of paramount importance, but even when cruising it makes life considerably less dramatic and requires much less physical effort and smaller winches.

Self-Tackers and Jib Booms

With a large mainsail supplying the bulk of the power, fitting a self-tacking jib also starts to make sense. If the sail has no overlap, why not have the track mounted transversely across the deck with a single sheet controlling the clew? This makes life really easy for short-handed crews and also means less deck gear and running rigging. The only time it creates a problem is when you want to **heave-to**, as the jib clew will automatically run down to the leeward end of the track. But this can easily be cured using a temporary **uphaul** to hold the sail up to windward when needed.

And if you have a self-tacking jib, why not fit a jib boom for improved sail-shape adjustment? There have been several experiments with carbon-fibre jib booms, mainly instigated by its originator, Gary Hoyt. The advantage of using a jib boom is that the draught of the sail can be altered, as with a **loose-footed** main, by adjusting the **outhaul**, whilst the sheet attaches to the boom and only alters the sail's angle of attack to the wind. This way the sail remains trimmed during a tack, with the sheet just used to move the boom in or out. The boom helps to keep the sail flat and the leech taut when **running** off the wind as well, but is notoriously difficult to '**prevent**' should you be sailing anywhere between a **broad reach** and a dead run. Another problem to be aware of is the

The staysail boom or jib boom offers additional control over the headsail shape

danger of standing on the foredeck if there is the slightest likelihood of the jib boom gybing, as it will fly across the deck extremely quickly at knee height with the obvious danger of knocking you overboard. Where the device has become more readily accepted is as a staysail boom on a cutter rig, where you need to keep the sail quite flat most of the time so as not to backwind the main headsail. The popular Island Packet range of cruising yachts utilises the system on all its cutter-rig boats with a good measure of success.

Cutters and Slutters

Cutters are generally masthead rigged in order to get a large enough area forward of the mast for the second forestay and sail. Traditionally, cutters have used **running backstays** (runners) to absorb the extra fore-and-aft loads put on the mast by the staysail, but with steeply, aft-swept spreaders the need for runners is significantly reduced – if not eliminated entirely.

Although a cutter rig allows more 'gear changes' and is a more balanced rig for offshore sailing, it does have its disadvantages when used for inshore cruising. Because of its small size the staysail is really only useful on its own as a heavy-weather sail, so many will just hoist the yankee in light airs. But being so high-cut, the yankee has a considerably smaller sail area when compared to a low-cut headsail such as the genoa, and is therefore underpowered. Should a genoa be hanked on, the driving power will be considerably increased, but tacking the large, overlapping sail around the inner forestay becomes problematic and the chafing on both the sail and sheets can be dramatic.

A variation on the sloop, sometimes called a 'slutter' rig (combination of sloop and cutter), is becoming more popular for long-term cruisers. The slutter has two forestays quite close together (a foot or so apart), rather than set halfway back from the headstay as with the inner stay on a cutter. This allows a working jib – possibly even a self-tacker – to be carried on the inner forestay whilst a larger genoa, **gennaker** or cruising chute is carried on the outer stay for reaching and running. Both sails are usually on **furlers** so that, should the need for tacking or gybing occur when the genoa is set, it can simply be rolled up before tacking, then unfurled again on

The 'slutter' rig has two headstays close to each other

the new course, avoiding any problems with pulling the sail around the inner forestay. Some yachts, such as Island Packets, fix the outer forestay to a short **bowsprit**, giving more room between the two forestays to help alleviate this problem.

Standing Rigging

The masts on early sloops were almost always supported by forward and aft shrouds, in addition to the upper (cap) shrouds taken to the masthead via the tips of the spreaders. In the case of cutter rigs, there were also detachable running backstays that were attached to the mast at the same point as the inner forestay, to stop the mast from flexing too much in the fore-and-aft direction when the staysail was set.

Sometimes sloop-rigged yachts had a shorter stay forward of the mast, known as the baby stay, which countered the pull from the two aft lower shrouds and meant that the forward lowers could be eliminated. This has evolved further since the introduction of spreaders that are swept aft at an angle of between 10–25 degrees. The old idea of rig triangulation using forward and aft lower shrouds, or aft shrouds and a baby stay, has been superseded in most contemporary designs by aft-swept spreaders and single, aft lower shrouds attached to the same **chain plates** as the cap shrouds. Dudley Dix comments, 'this results in a more stable mast without excessive fore-and-aft rigging or mast section. The shrouds pull the centre of the mast aft, while at the same time backstay tension and mast compression loads are working to bend the mast forward, applying opposite loads to those of the shrouds. The result is that the mast can bend only as far as the shroud lengths will allow.'

A further advantage of sweeping the spreaders aft is that it narrows the effective shroud-base distance without reducing the crucial shroud angle to the masthead. This allows the **jib-sheet** track to be placed further inboard, which in turn enables a tighter sheet-ing angle to be set for better windward pointing.

However, if the hull isn't shaped accord-ingly and her shoulders are too broad (common with high-volume cruis-ing boats) to slice through the seas, tightening the sheets will simply lead to an increase in leeway, even if she appears to be pointing better. For this reason designers have moved towards narrow, flared forward sections and plumb stems to complement the rig. As renowned naval architect Bob Perry says, 'A balance must be struck between hull form, beam, sheeting angles, keel geometry and draft. There is no point

Sweeping the spreaders aft removes the need for runners or forward shrouds

in having reduced draft and maximized beam (fullness) and then putting it together with a close sheeting angle.'

The absence of forward lower shrouds also means that the jib can overlap the mast a little without fouling its leech on the forward shrouds. In many cases this increased headsail area does away with the need for a genoa entirely.

There is, however, a downside to overly aft-swept spreaders in that a narrower shroud base can result in less lateral support to the mast. Rather than extending the length of the spreaders, this is usually countered by adding a second spreader and more of the lateral bending loads are kept within the rig itself by using intermediate (**jumper**) shrouds, rather than transferring the stress onto the hull and deck. If the shroud base is made narrower and the number of spreaders stays the same then the shrouds have to pull harder to keep the mast standing. They also pull more downwards if the shroud angle reduces, so mast compression loads increase dramatically. You have to keep a reasonable angle on the shrouds so that they pull sidewards effectively, which can only be done by increasing the number of spreaders. In doing this, the spreaders are closer together, so the distances between the support points on the mast are reduced. This allows smaller mast sections to be used that are lighter and have less windage – i.e. less cross-sectional area shadowing the mainsail from the wind.

Another notable drawback of having aft-swept spreaders is that when you're running downwind they restrict how far the boom can be let out before the mainsail starts chafing on them. This often means you are forced to constantly gybe downwind, rather than set the boat up on a **dead run**

with the boom almost at right-angles to the centre-line. A compromise is to sweep the spreaders no more than 10 degrees or so aft, then add detachable running backstays for use in the event of a real blow, but this is rarely the practice on cruising boats designed for short-handed sailing, as runners add cost and give the limited crew more tackle and lines to worry about.

A further consideration high on the production boatbuilder's priority list is the reduction in costs involved when there is only one pair of chain plates to build into the hull. Considerable savings are made through reduced hull strengthening and rigging hardware.

Continuous or discontinuous?

In the early years of the more modern rig, cap shrouds were almost always 'continuous' – i.e.

Spreader ends showing the arrangement for discontinuous standing rigging

they were made from a single piece of wire fed through guides at the end of the spreaders before terminating on deck-mounted chain plates. More recently, however, 'discontinuous' standing rigging has become popular with designers, whereby the cap shroud from the masthead is terminated on a load plate fixed to the end of the spreader and a second wire is taken from the lower half of the plate to the chain plate or, in the case where the mast has multiple spreaders, the end plate on the next spreader down, and so on. This arrangement overcomes the huge stresses involved in keeping the correct tension on very long cap shrouds and avoids the age-old problem of wear at the spreader tips.

Wire or rod?

Virtually all cruising yachts built after the 1960s utilised stainless steel, multi-stranded wire for their standing rigging, constructed either in a 1 × 19 (strands) or 7 × 7 format (seven groups of seven wires twisted together). Towards the end of the century, however, the use of solid rod for standing rigging started to become increasingly popular, although mainly for racing boats and very large cruising yachts.

Rod rigging displays less stretch and creep than multi-stranded wire, which allows the rig to hold onto its correct shape and tension for longer, but it is also considerably less forgiving. When setting up the rig and tensioning the stays with rod, it is vital to be able to measure the tension on each component accurately, or you will find you can easily damage the hull. Cranking down a rod rig too far has been known to bend a hull into a banana shape, forcing the mast down into the boat and even lifting the hull away from the keel at the fore and aft extremes. Designers often incorporate electronic load cells into the rod itself to keep the rigger or skipper accurately informed of the rig loads at all times.

Rod rigging is more often used on superyachts or racing boats

Rod rigging is more difficult to work with and attaching terminals is definitely best left to the professionals. It goes without saying, then, that it is also significantly more expensive to install and maintain.

Bergstrom and Ridder rig

The swept spreader rig was taken a step further with the Bergstrom and Ridder (B & R) deck-stepped design. It was developed initially for small, lightweight racing yachts, but was soon adopted elsewhere for both racing and cruising yachts. The idea behind the B & R rig, thought to be quite radical when it first appeared, is to contain as much of the rigging loads within the mast and rig structure itself to avoid loading the hull and deck. To this end, the spreaders are swept back a massive 30 degrees or so off the beam line, giving such effective forward and aft support to the mast that it can even eliminate the need for a backstay. The B & R rig allows the use of a lighter, more aerodynamic mast section than previously deployed as the mast is prevented from bending in its lateral plane by multiple spreaders with double-diagonal, discontinuous jumpers.

Its Swedish inventors, Lars Bergstrom and Sven Ridder first started incorporating the principles into designer Paul Lindenberg's early, lightweight racing boats circa 1975. In 1984 the sailing world started to notice the success of the B & R rig after the Lindenberg-designed, 60 ft *Hunter's Child* won the OSTAR using the rig. Owner, Warren Luhrs, went on to form Hunter Boats USA (marketed as Legend yachts in the UK to avoid confusion with British Hunter Yachts), which has continued to utilise the B & R rig throughout its range right up to its current range of sporty cruising yachts from 30–50 ft. Hunter's rigs are now mostly fractional, except for some of the largest yachts in its range, and none have a backstay. This lack of backstay enables more **roach** to be put in the leech of the mainsail, increasing the sail area

The B & R rig on a Hunter Legend has no backstay

for better downwind performance. In a later development of the rig, Hunter reinforced the mast on its larger yachts by incorporating rigid struts between the chain plates to a point on the mast just above the **gooseneck**. The purpose of these was to distribute the compression loads on the mast between the chain plates and the mast step, significantly reducing the amount of reinforcement required to the superstructure to take the down-force on the mast.

Thousands of these rigs have now been built and tested in all sea and weather conditions without failure, so the concept has been proven beyond all doubt. However, in common with all aft-swept spreader rigs it does suffer from limited boom movement downwind. Another problem that arises when beating in strong winds with a backstay-less rig is forestay sag. To some degree a powerful boom **vang** helps, but it will always suffer from this problem, which will affect its pointing ability to windward. Earlier masthead versions of the rig had backstays, but tightening them down to create more mast bend can loosen the upper shrouds, making the mast more prone to lateral flex – not a desired condition when beating hard!

The shrouds in the B & R rig are designed specifically to take over from the backstay and are tensioned to a fixed degree, not tweaked as you sail. Some forestay sag can be eliminated by increasing the tension in the shrouds (particularly the uppers), but this is more of a one-off tuning adjustment rather than something that can be carried out under sail. An adjuster on the forestay would also help, but with roller furling fitted would just complicate matters.

The rigs must be set up and tuned very precisely when first stepped. When tuning the rig the upper shrouds are tensioned first. As the tension on the shrouds increases, the middle of the mast bows forward, aided by the swept spreaders. When the desired amount of pre-bend is achieved, the lowers are tensioned to stiffen the middle of the mast. The tension in the lowers is usually around half that in the uppers.

Masts – keel or deck stepped?

Keel-stepped masts are inherently better supported than deck-stepped ones because a large portion of the root is firmly retained by the suitably reinforced deck around the partners (the hole in the cabin top) and they don't put any compression loads on the coachroof. However, they are trickier to set up due to the need for wedges between the mast and partners and also they have a tendency to allow water to run down through the hole in the coachroof, despite having a gaiter attached. Also, should the rig fail completely and the mast is lost overboard, it can create considerable damage to the superstructure.

Because of the tendency towards keeping the rig loads within the mast these days, most new production offshore cruising yachts have deck-stepped masts with an internal post or lintel transferring the compression loads down to the keel. A deck-stepped mast is also easily adjusted for rake angle and is much easier to step/unstep when required.

A typical keel-stepped mast showing the gaiter, but deck-stepped masts are more common these days

Reefing Systems

In sailing, as with almost every other aspect of our lives these days, technology is marching ahead and new developments, intended to make our lives easier, are continually being introduced. With the average size of cruising boats getting bigger all the time, much thought is being put into the design of the sail handling systems in order to make them simpler, safer and less physically draining to operate.

With the exception of **GPS**, probably the most useful aid ever devised for cruising yachties is the roller furling headsail. This simple device has undoubtedly encouraged more husband and wife teams to set sail than any other piece of equipment on the boat, given the need with the older-style hanked-on sails for a crewmember to go forward and change the headsail when the going got tough. Now we are demanding the same ease of handling for the mainsail, but which method is the best in terms of performance, safety and cost?

In-mast versus in-boom furling

Mainsail furling was first used some 50 years ago to offer the yachtsman an easier way of taming the main than **slab reefing** with countless reefing pennants that had to be tied in. The early versions were simply rotating booms, but they caused problems by spoiling the shape of the sail and preventing a proper **kicking strap** from being fitted.

A typical headsail furler for reefing the genoa of a cruising yacht

Some older boats were modified by retro-fitting external mainsail furlers

In the late 1970s, furling devices that acted like roller blinds bolted onto the back of the mast were introduced. Whilst these overcame some of the problems of the rolling boom, they did little to overcome the sail shape problem and often compromised stability and reliability by adding considerable weight aloft and jamming when the sail wasn't kept under the correct tension and angle when furling it back in.

However, once the idea had started to catch on, dedicated spars were manufactured that integrated the furling mechanism into the mast section itself, eliminating many of the tiresome mechanical problems of the retro-fit, behind-mast furler. More recently, the use of flexible vertical battens in these sails has improved sail shape and performance markedly and we are nearing the point when, for cruising yachtsmen, the small sacrifice in performance over a standard, slab-reefing mainsail might make them well worth considering.

Both in-boom and in-mast mainsail furling methods have been used fairly extensively for a good many years now and, though there have been many improvements during that time, both still have a few hurdles to overcome before they achieve the reliability and performance of headsail furlers.

In-mast reefing

Virtually every mast manufacturer now produces an in-mast furling model for all sizes of cruising yacht and by incorporating sail design enhancements such as vertical battening, they have become considerably more efficient.

However good the latest designs of in-mast or behind-mast reefing systems might be, though, they will never be able to overcome certain disadvantages such as:

- The inability to lower a partly furled sail in the event of the mechanism jamming up – although nowhere near so common these days, it used to happen quite frequently with early designs and often forced a crewmember to climb up the mast to cut the sail down in high winds and big seas.
- The additional, unwanted weight aloft of the furled sail, internal mandrel and thicker spar when reefed can adversely affect the stability of a light displacement yacht in heavy weather.
- The extra windage high up in the partly reefed sail, caused by the centre of effort remaining too high, means that less sail can be left out. With the more traditional slab-reefing the centre of wind effort comes down closer to the waterline as the sail is lowered, resulting in a reduced heeling effect.
- The sail has to be cut flatter to avoid any extra material from bunching up and jamming in the mechanism – resulting in a poorer performance in light airs.
- Unless vertical battens are incorporated, in-mast furling mainsails are commonly made with no roach (the area of sail beyond a straight line between the **head** and clew), again reducing the area of sail available for light airs.
- It is not possible to adjust the tension in the backstay for **sail trimming** purposes.

Overall, the limitation on its shape undoubtedly compromises a yacht's sailing performance to a noticeable degree, unless all the latest advances are incorporated.

In return for this lessening of windward performance and loss of stability when reefed, in-mast furling does make life significantly easier on board for a short-handed or inexperienced crew, which, given the ease with which the mainsail can be shortened from the cockpit, is less likely to find itself with too much sail up when conditions deteriorate.

In-boom furling

One possible answer to these problems would appear to be to furl the mainsail inside the boom, something that has been experimented with since Eric Hiscock toyed with worm gear-driven spindles on his beloved *Wanderer/s*. The main problem with earlier, rotating boom designs was that they prevented fittings such as kicking straps (aka vangs) to be attached to the underside

The in-mast furler is now de rigueur *on many coastal cruising yachts*

of the boom. Also, the **mainsheet** had to be fastened to a rotating boom end fitting – something that used to wear out or jam up with monotonous regularity.

Since then in-boom furling has become available, firstly only on larger yachts, but now on medium-size cruising yachts as well. After worldwide dialogue between sail and spar makers, modern versions of the boom furler are at last becoming a reliable alternative to in-mast furling. The system most commonly used is similar in principle to in-mast designs, only mounted horizontally. That is, the boom becomes a hollow U-shaped shell and a rotating mandrel, around which the sail is furled, is suspended on bearings at each end.

The boom itself must be supported by a rigid vang, not only to retain control over the sail's draught as normal, but also to ensure that the optimum angle between the mast and boom is maintained in order to be able to reef the sail without it bunching up. As with foam-luffed furling

In-boom mainsail furlers are fairly rare on small craft and they are expensive

headsails, a foam foot can be sewn into the mainsail to take up the draught, allowing the sail to furl evenly along the roller. This means that, as with conventional slab reefing, both the weight and the centre of effort of the sail are lowered when you need it most. Furthermore, the sail can be dropped quickly and tied around the boom in the conventional way should the whole thing grind to a halt.

Modern versions of the furling boom, though much improved on the original, still have a few drawbacks over simple slab reefing:

- Firstly, despite the mechanism allowing a positive roach and full-length horizontal battens to be used, it is not possible to fit cars to the batten ends, so the full pressure along the batten is taken up by the batten pockets, which have to be greatly reinforced and tend to wear out rapidly from chafing against the mast. The lack of cars also makes the sail harder to hoist, so the mechanism is usually electrically driven on boats over 40 ft or so. Saying this, I have seen some ingenious systems developed for superyachts whereby the sail has clips attached to the luff, which 'picks up' a car on its ascent. But these are phenomenally expensive and too sophisticated for general cruising use on small boats.

- Secondly, some sail shape has to be sacrificed to prevent the sail becoming a total bag of laundry when furled more than a few turns. Although most incorporate a mandrel that is thicker in the centre than at the ends, thereby hauling the centre part of the sail in faster than the rest, some sacrifice to sail shape and reduction in draught is inevitable. Also, without reefing lines to outhaul the sail at the crucial reefing points, the sail tends to creep forward on the mandrel, increasing the bagginess still further.

- Thirdly, there is the problem of containing the greater mass of sail forward on the mandrel without it clogging up the boom at the forward end. This, combined with not being able to outhaul the sail when furling, often causes the sail to jam inside the boom at the forward end. Attempts to fix these problems include the use of larger boom shells tapering towards the aft end, which are wide open at the mast end to intentionally allow the sail to move forward. In doing so, the luff winds in a slight spiral, rather than lying on top of itself, thereby taking up less room.

- Finally, the boom to mast angle is critical to the furling in order to stop the sail from bunching up and the battens spiralling around the mandrel. This is partly overcome by the use of gas-sprung kickers that still allow the boom to be hardened down to flatten the sail in strong winds, but always return to the optimum position that creates the ideal furling angle when released.

In conclusion, the in-mast furling systems seem to be better developed at the moment than in-boom mechanisms, and appear on almost every boatbuilder's options list for little extra cost. In fact, some even offer them as standard kit, with fully battened slab reefing mainsail being the extra cost option.

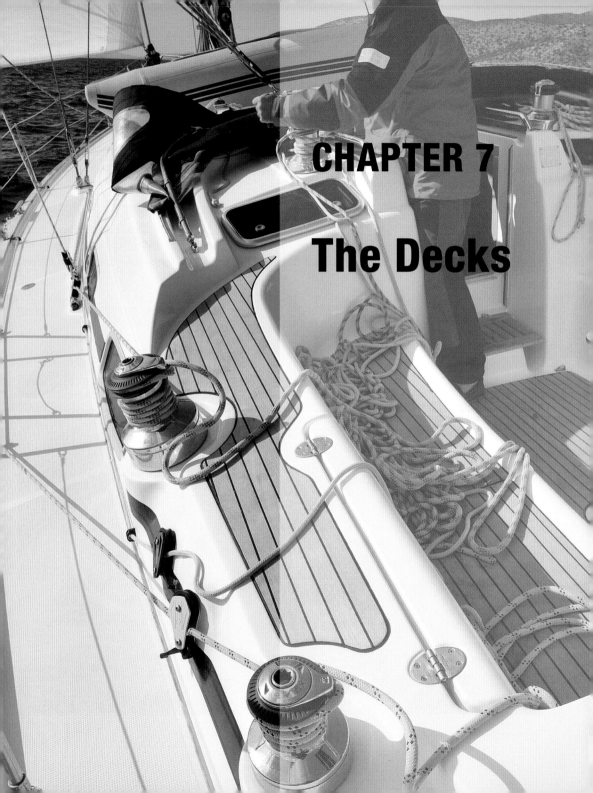

CHAPTER 7

The Decks

Cockpit

Today's cruising yachts have a far more generous beam than boats of 20–30 years ago. This extra width is usually carried almost all the way aft, allowing designers to squeeze in two generously proportioned double aft-cabins, as well as making the cockpit a good size for entertaining in port or at anchor.

This doesn't necessarily create an ideal work pit for the crew, however, as a wide cockpit means there's further to fall when the yacht heels under sail. As usual, it's all down to a compromise. There needs to be something to grab hold of to prevent you being thrown from one side to the other – a cockpit table, foot bar or stout handrail on the top of the binnacle – and the coamings

This handy grabrail and foot bar converts into a cockpit table when needed

(seat backs) need to be high enough to keep out the waves, as well as being smoothly contoured and at the right angle for comfortable back support when you're seated.

Twin steering positions might seem an extravagance in a yacht less than 50 ft long, but they do create an easy passage aft and separate the helmsman from the rest of the cockpit. This can be mutually beneficial when cruising with non-sailing passengers who have a habit of getting in the way!

The design and practicality of a cruising yacht's stern has taken on more importance these days, with the popularity of swimming and other watersports such as windsurfing, kayaking and diving. In older yachts the transom usually overhangs, so that getting in and out of the sea is often quite a strenuous task.

With the advent of the retroussé transom (more commonly known as 'sugar scoops') in the late 1970s, it became considerably easier to mount a decent boarding ladder. A little later,

Will the drop-down boarding platform become more popular than the sugar scoop?

designers realised that they could put a moulded step or two into the transom, and before long the lower step evolved into a swimming and boarding platform. Nowadays, transoms appear to be reverting to the more upright design, but one that incorporates a drop-down platform. This creates much more useable space below and has the added benefit of cutting down on marina costs – after all, who wants to pay £3 per metre, per night, for a set of boarding steps?

Almost every new cruising yacht built today is labelled as some form of cruiser-racer by its developers, so every boat must be able to offer an exciting sailing performance as well as providing lavish living quarters with all the trappings of a waterside apartment. I'm not criticising this in principle, as all yachtsmen like to get the best out of their boat under way, but it can lead to a few rather edgy compromises.

Most racing yachts have open transoms these days, which in many ways is perfectly fine. You couldn't get a more effective cockpit drain than an open transom and it allows designers to move the helm further aft, giving the crew more room in which to work the sails. An open transom is often combined with a large wheel that spans the entire width of the cockpit and separates the helmsman from his working crew. But in cruising terms, this is an impractical set up. It isolates the helmsman and makes access to the stern difficult. Furthermore, it forces him to climb up and over the seating to go forward, increasing the risk of losing his balance and being tipped overboard in the process.

Some designers have tried to compensate by providing removable seats and detachable **quarter** lockers that can be left ashore when racing and firmly attached for cruising. Personally, if I were planning an open-ocean passage, I'd far rather have a solid transom to prevent the risk of being swamped by a steep following sea.

In almost every yacht built prior to 1980 you will find a substantial bridge deck – that is a tall step between the cockpit and **companionway** steps – to ensure that no water can enter below should a large wave engulf the cockpit. This was absolutely essential in the days before self-draining cockpits, but nowadays the bridge deck seems to have been replaced by a less tall and intrusive threshold, often supplemented by a lower washboard that can be left in place during heavy weather conditions.

It is also quite common, on these early boats, to have the mainsheet **traveller** track running along the top of this bridge deck. Though this is often convenient for trimming the mainsail, especially in **tiller**-steered yachts, it can prove a little scary when cruising – especially with young children whose fingers tend to stray into compromising areas in the blink of an eye.

Some yachts have the mainsheet just forward of the wheel or, in tiller-steered craft, across the centre of the cockpit well. Whilst this might be the ideal position from a purely engineering point of view – enabling a straight up-and-down mainsheet with multi-block sheet control and simple

An open transom like this isn't ideal for offshore following seas

jammer – it is inconvenient when cruising with family or guests. For this reason, on post-1980 yachts the mainsheet track is more commonly positioned just forward of the main hatchway, where it is out of the way of probing fingers. Unfortunately, this usually means it is halfway along the boom, which makes it a heavy line to trim in anything resembling a good breeze. Worse still, by taking the mainsheet forward along the underside of the boom to the mast, down to the mast step, around a couple of turning blocks, then back via a rope clutch to a coachroof-mounted halyard winch, the friction created is so great as to require the obligatory use of a winch for trimming the main.

When sailing short-handed, as most cruising crews will be, it can be helpful if the primary (headsail) winches are close by the helm, so that the helmsman can trim the jib without leaving the wheel. Although in masthead yachts with large, overlapping genoas, tacking single-handedly can be a real handful unless you have a smart autopilot. Also, it's not much use being able to trim the headsail from the helm if the mainsheet is out of reach!

Having the mainsheet traveller mounted on the bridge deck can be dangerous

The popularity of twin helms these days has brought the double-ended mainsheet, commonly called the German mainsheet, to the fore. This system allows the helmsman to use either the **port** or **starboard** winch close to the helm to adjust the mainsheet, meaning it can be done from whichever wheel is currently in use. However the friction created by taking the mainsheet from the coachroof, forward to the mast and then back to the helm means that, though it can be let off easily, it needs to be winched to sheet it back in. Fine if you have electric winches, but pretty slow if you want to be able to 'play' the mainsheet in a blow.

Stowage

On a cruising yacht it is essential to have good deck stowage for all the gear you're likely to need on board, so deep cockpit or quarter lockers are essential. But it's no use having cavernous,

Goodsize cockpit lockers are essential on a cruising yacht

full-depth lockers unless they have compartments or some form of shelving for smaller items. Otherwise everything ends up in a heap at the bottom. Look out for steps, lights and smaller divisions, which will make a boat's lockers far more user-friendly. It's good to have one with a large opening as well, for bulky items such as an inflatable dinghy, outboard engine, **kedge** anchor and jerry-cans. A dedicated sail locker is a real bonus, as is somewhere to stow the washboards.

Shelter

If you're planning to cruise rather than race you'll probably be glad of a good sprayhood for the cockpit. You need something to duck behind when it gets a little wet on deck. If one comes with the boat, do take a good look at it as this is an expensive item to replace.

Some boats will have a full cockpit tent, which is even better if you're out in all weathers. Although you can't sail with it up, it makes a wonderful 'conservatory' at anchor and really can extend your sailing season – particularly if you have heating on board. On long cruises it is often frustrating to find yourself sitting down below in the gloom when the weather's unpleasant. It's so much nicer to be able to see what's going on around you. Plus, a cockpit tent can make a great place to put up any extra guests for the night too!

Finally, give some thought to practical creature comforts. I find I can't do without some form of cup holder, so that, should I need to put in a sudden tack, I can safely slot my mug of tea into a holder and concentrate on the boat. Such extras won't make a boat a 'must buy' but the cost in time and money of installing them should be considered when evaluating a potential purchase.

Cockpit tents add living space and can extend your sailing season

Decks

With boats becoming so beamy there's little excuse not to carry the side decks all the way aft, or at least to a point where it's possible to enter and leave the cockpit safely. Some models, however, still insist on widening the coachroof as far as possible to enlarge the volume of the interior. Light and airy might be the buzzwords of today when it comes to interiors, but side decks that are wide enough to walk along unhindered are an important safety factor that shouldn't be adversely affected by the desire for a spacious cabin.

Guardrails can be more of a hindrance than a help sometimes – especially when the decks are narrow. Many are not high enough and can cause you to flip overboard rather than save you. Either way, I would never recommend relying on them as a handhold – particularly as many of them are simply held in place with self-tapping screws rather than being bolted right through the deck. **Stanchion** bases are also one of the most likely causes of leaks, as they are often grabbed when boarding or fending off another boat. Some skippers even use them to tie mooring lines to, in the absence of midship **cleats**. I never rely on them for anything – in fact I'd be happier not to have them at all, like boats built in the early 20th Century, as it soon teaches you a thing or two about holding on!

A good **toe rail** (better still a raised **gunwale**) and **jackstays** are a much safer solution than relying on the guardrails to keep you on board, especially if you adhere to the wise old edict, 'one hand for the boat and one hand for you'. So, unless there are more compelling reasons to buy a particular yacht, I would do my best to avoid one that has no side decks (yes, they do exist), or ones that are extremely narrow and cluttered with rigging, jib tracks etc.

One of the most important areas of the deck is the foredeck – particularly on a cruising yacht, as you are likely to be spending a good deal of time at anchor or picking up buoys. A good, sturdy stemhead fitting is vital – preferably one with double **bow-rollers** so you can put out a second anchor in a real blow.

Then you'll need a deep chain locker big enough to hold at least three times the boat's length in chain. Older boats usually just have a chain pipe down which the chain is fed into a locker below, but this can often cause problems with it jamming up somewhere along the way. It is much easier to sort out if the locker is accessible from the deck.

Modern stemhead fittings are also designed for self-stowing anchors, which alleviates the need to lash the anchor on deck when inshore cruising. Many newer yachts will also have an electric **windlass** (anchor chain winch), which may be deck-mounted, or better still, set on a plinth inside the chain locker leaving the foredeck clear of obstructions. Although an electric windlass is often classed as an expensive luxury item, if you are cruising for any length of time I think it's a piece of equipment you wouldn't want to do without. If the anchor holds first time, then fine. But try hauling in 100 ft of heavy chain several times in quick succession while attempting to get the anchor to set in an area of variable holding – then you'll see what I mean.

Raised gunwales give your feet something to grip onto when seriously heeled

Another essential deck fitting is the humble mooring cleat. Put simply – you can't have enough of them! They need to be a good size and bolted through the deck with substantial backing plates as they will take a hammering over the years. If you buy a boat that doesn't have a pair amid-ships, and you plan to sail single-handed or two-up, then I would suggest you fit a pair as soon as possible. Sometimes, if you have a perforated toe rail, you can buy purpose-designed bolt-on cleats. Otherwise, you'll need to mount them on the deck and bolt through to below. You might also need to cut away a little of the toe rail to fit an extra fairlead as well. Some owners simply fit a fairlead amidships, then take the warps along the deck to another strong point such as a winch.

There is much to recommend a sturdy Sampson post in the middle of the foredeck – something that was very common on older craft that often needed towing. It provides a fantastic strong point for a multitude of uses and is ideal should you ever need a tow back to port one day!

Bolt-on cleats – a handy option for yachts with aluminium toe rails

A good non-slip surface is also essential on a yacht and this can come in a variety of guises. At one time, people who couldn't afford teak-planked decks simply painted them with a mixture of paint and sand, which is, in fact, pretty effective. Nowadays, most GRP yachts have a non-slip tread pattern moulded into the deck's upper surface. Some are more effective than others, but it's easy to boost the important areas if needed by gluing on strips of non-slip covering. This is quite common for areas such as slippery forehatches or sloping edges of the coachroof.

One of the most popular deck coverings in the 1970–80s was Treadmaster – a patterned, rubberised material that was firmly glued to the deck. Still available today, it works extremely well, often lasting 20 years or more, but it's a devil to get off and replace without damaging the deck underneath – especially if the boat has plywood decks. It can also be used to hide a multitude of sins, however, so do check it very carefully and any surrounding uncovered areas. If you suspect the deck might be waterlogged underneath it, get the surveyor to put his moisture meter over it.

Handrails are essential and the longer the better. I've never understood why some yacht builders provide them only a couple of feet in length, when ideally they should run the entire length of the coachroof. Also, if the boat has all its reefing lines and controls at the mast, which is common for older boats, then steel 'granny bars' around the mast base are a real boon. They may look like a pair of Zimmer frames screwed to the deck but they provide support when reefing and somewhere to tie off any spare lines away from the mast.

One final item that all cruisers will need to consider when sailing away from their home port for a few days is the ubiquitous dinghy, or tender. Most yachts use inflatables these days, as they can be deflated and stowed in a smallish cockpit locker when not in use. But for long-term cruisers,

'Granny bars' provide excellent support when reefing down in heavy weather

Davits on the stern are convenient for stowing the dinghy but may also restrict all-round vision

the tender and outboard is as essential as your car is at home and needs to be immediately available for a multitude of tasks. For this reason owners often fit davits – steel 'jibs' to the afterdeck, on which the tender is suspended out of the water. If properly mounted these will support the dinghy when you're sailing, but they can restrict your all-round vision on smaller yachts and create extra windage. Some modern yachts are being designed with large storage areas accessible from the transom. Known as 'garages', they become large enough to stow an RIB on vessels of 60 ft or more. Outboard motors are most easily stowed on a proper mount attached to the stern rail, or **pushpit** as it is sometimes called. With today's engines getting larger and heavier, it's also best to fit another small jib for raising and lowering it onto the dinghy transom. Otherwise you'll quite likely hurt your back, or even drop the outboard in the water. Designs such as the Ovni feature 'goal posts' on the stern that act as mounts for such jibs and davits as well as aerials, solar panels and wind generators.

CHAPTER 8

Accommodation

Although probably far more of a concern to liveaboards than those day-sailing or weekending, a yacht's accommodation can be the make-or-break of the sailing experience. It might not be so vital in small speedboats or RIBs, as they can get you home in the blink of an eye, but in small sailing boats it's not only very nice to have some respite from the elements, but also it can prevent crews getting seasick, or even hypothermic.

Small day-boats of 12–16ft are unlikely to stray far enough away from a safe haven, so they might not need any fixed cabin protection. Often a simple canvas boom-tent will provide shelter from the rain while you're having a cuppa at anchor. But it's nice to have somewhere dry to stow a few items of warm clothing as well as some provisions if you're out for the whole day.

A simple boom tent can offer shelter and warmth

Most small cruisers will have some form of cabin, although those under 25ft long will rarely offer full standing-headroom. This isn't a big problem if you're only staying on board for a night or two. Most of the time when you're below decks you'll be eating, sitting or sleeping – all of which can be done without standing upright. It can sometimes get a little tricky when cooking, but small-boat catering will rarely involve too much fancy work. In fact most of the time you'll probably just be boiling a kettle or reheating a pre-cooked stew or casserole in a single pot.

Dressing with limited headroom can be a little awkward, but you can usually manage without too much trouble in the seated position. Use of the loo means sitting I'm afraid chaps, but in my experience this makes a lot of sense anyway and avoids any mopping!

If you plan to spend more than one or two nights on board you might want to consider the interior a little more seriously and I'd definitely look out for a boat with at least 5ft of headroom. Stooping the entire time you're down below can cause some pretty awful back and neck aches. Ideally, there should be at least one place in the boat you can stand upright to stretch the spine. In my current boat, a Jaguar 27, I can just stand fully upright (6ft) beneath the large, main hatch, which is a blessing after a while, although for the rest of the time 5 ft 9 in is just fine. One previous boat had just 5 ft 4 in headroom, which was tolerable – at least for a few days.

Accommodation

There's more to yacht interiors than just headroom, however. Cruising for extended periods requires some form of 'normality' in the facilities. I'm not saying it should provide all the facilities you have at home, but you should feel comfortable and able to carry out most daily tasks such as washing, cooking, dressing and eating without it causing too much stress. The areas that can cause the most frustration are the galley, heads and navigation station. Sleeping quarters can be pretty basic really – so long as the bunks are long enough and reasonably well padded, you shouldn't have too much trouble getting to sleep.

Keeping gas bottles in a non-draining cockpit locker can be dangerous

Galley

If you plan to spend a week or more on board the galley will be important – unless you're too lazy to cook or wealthy enough to eat out each night. Most boats use bottled gas for cooking, which isn't ideal from a safety point of view, but it is convenient and closely simulates cooking at home. There are a few things to look out for, though, with gas installations. They really should meet the latest gas safety regulations, so I'd recommend anyone who buys a used boat over three years old to have the gas system thoroughly checked for compliance. Leaks can, and often do occur, and the results can be fatal. Gas is heavier than air, so any leaks from chafed pipes or poor joints and the gas will sink to the lowest point in the boat – well before you get to smell it. Then it only takes one spark – from an automatic bilge-pump switch, say, and *Ker-boom*!

On every boat I've owned that carried bottled gas for cooking, I have fitted a gas detection system plus a solenoid shut-off valve back at the bottle that I can activate from a switch below. This saves me going out in the rain to turn it off at the bottle. However, being rather paranoid about gas leaks (I've seen the unpleasant result of a gas explosion on a yacht) I still make sure it is turned off at the bottle before I turn in at night or leave the boat unattended.

Another important detail is to ensure that all the gas bottles are installed in a sealed locker, which contains nothing else and is drained overboard from its lowest point. Many old boats don't have this facility, so often bottles are kept in open, undrained cockpit lockers. I've also seen quite a few gas-bottle lockers that drain into the cockpit – presumably on the assumption that any leaking gas will find its way down the cockpit drains. However, unless these drains exit the hull above the waterline, which is unlikely, this just doesn't work. The leaked gas simply collects in the drainpipes, unable to escape because the sea effectively seals the ends of the pipes. The danger is obvious.

If you plan to spend most nights in a marina, then carrying big bottles of gas on board may be unnecessary. On my boat I've got an electric induction ring with adjustable power settings, a small microwave oven with electric grill, and I keep a simple, one-ring portable gas cooker that takes small cartridges of gas directly inserted in the side, for when I'm at anchor. When I'm in a marina with power, I cook up several good main meals, which I then seal up in plastic containers ready to be briefly reheated using the portable camping cooker when I'm away from shore power. I also use my electric kettle to fill a large vacuum flask, so that I don't need to heat water on the portable cooker for at least a day.

The biggest problem with any onboard catering is the fridge. You really can't do without one if you're going to be aboard for more than a few days, especially in the Med or the Caribbean. But of all the boat's equipment the fridge is probably the most voracious consumer of electrical energy. It's not so much of a problem if you have bags of battery power and are within reach of shore power every other night, but if you have limited energy reserves, a fridge can often be more of a hindrance than a help.

Sometimes, particularly with smaller boats and day-sailers, it's better just to get yourself a very well insulated cool-box and fill it with frozen water bottles or large chunks of ice. I once saw a

A seriously large fridge/freezer on a blue water cruising yacht

boat with two cool-boxes – one large enough to fit the other inside – plus space between the two to fill with foam. A double-insulated box like this can keep ice for a couple of days at least. On my last boat, a 35 ft blue water cruising yacht, I had the full works – a compressor-driven fridge with a water-cooled condenser on the keel and four large domestic batteries – but it was still the most power hungry device I had on board and it was a pain to have to run the engine in an anchorage every day, just to keep the fridge cold.

Heads and showers

It's nice to have proper toilet facilities on board when you're cruising. A portable loo might be okay for the day, but if you're sleeping aboard you'll probably want something a little better. Privacy is pretty rare on a small boat, but it's a real bonus to have a separate heads compartment – especially if it has a washbasin as well. Of course cruising yachts will have much larger heads – usually with a hot shower as well – but this isn't necessary for day trips or marina hopping, as there will always be facilities ashore.

Nowadays long-term cruising yachts need to have a waste holding tank that can be discharged well out to sea, or better still pumped out in a marina. Loo pumps on board are usually fitted with diverter valves so you can choose between pumping out directly into the sea or diverting the waste into the holding tank to be emptied out later. Some European countries also insist that you don't discharge grey water (shower or washing-up water) either as it can damage the marine habitat. In this case, you need to find a way of storing grey water as well, or at least have a large enough waste holding tank to take both.

Most modern yachts over 25 ft long have a hot water shower in the heads

Although a shower isn't likely to be deemed essential for coastal or day-sailing, hot water is pretty straightforward to provide if you have an inboard engine. Most modern cruising yachts will usually have a pressurised water system anyway, so the likelihood of having a shower on board a production yacht is now very high. However, it's not really considered reasonable or eco-friendly to use your onboard shower in a marina if facilities are provided ashore. The less detergent discharged into harbours and rivers the better really. Biologically friendly soaps and shampoos are widely available, though, so it's well worth looking out for them for onboard use.

Navigation station

Having spent a large chunk of my life living on board cruising boats, I'm not sure I could do without a proper navigation area. Yes, I know the modern way of navigating is by chart plotter at the helm, but I think it's a good idea to have one area of the boat that the skipper can call his own. In fact I'm quite territorial about mine – I won't let it be used as an extra galley worktop or allow people to put drinks on the chart table. To me, this is the skipper's workstation; the place where the serious work of passage planning, log keeping and fix plotting is done. I also think it's good to have a central point of information – a place where all the sensors, receivers, monitors, indicators and controls all terminate – somewhere I can go to check that all's well aboard, both in the navigational sense and with the well-being of the vessel itself.

A dedicated navigation station is essential in an offshore cruising yacht

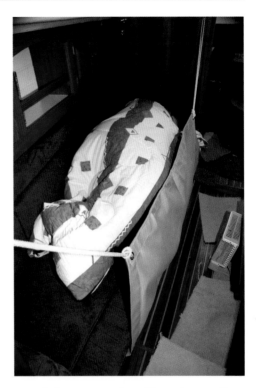

Lee cloths prevent the off-watch crew from falling out of their bunks

The navigation station is a place where you can instantly find all the necessary pilot books, almanacs and tide tables, and somewhere I can always glance at the instruments when not on deck.

It also needs to have enough dedicated space for a good wad of spare charts, at least a dozen pilot guides and books, plus a variety of manual plotting instruments, pencils, erasers etc.

On a small day-boat this might well be considered unnecessary, but even on an open day-sailing boat I like to have a portable nav kit – usually just a watertight Ziploc bag containing a chart, pilot book and plotting instruments.

Sleeping cabins

Coastal cruisers under 20 ft or so are unlikely to have separate sleeping quarters, but most will have a couple of settee berths in the **saloon** that are usually quite comfortable – if a little short and narrow. Voluminous cabins with wide double berths are fine at anchor or in a marina, but not often much use at sea unless you can fit them with a **lee cloth** (a canvas panel designed to stop you rolling out of a bunk). Worse than wide double berths at sea are those that are placed transversely across the boat. The motion in these at sea is a lot worse than in a bunk that is parallel to the yacht's centreline.

The ideal sea berth, particularly for those prone to seasickness, is one that is as close to the keel as possible, which usually means one of the saloon settees with a lee cloth attached. This is where the motion of the boat will be the least noticeable. Bunks in the forepeak are probably the worst, as they are the most affected by slamming and wave motion, and they will certainly be the noisiest under way.

A fully enclosed cockpit tent can make all the difference on a chilly autumn eve

Large aft cabins can be very comfortable when you're not at sea, but if the yacht has a wide, shallow stern there will often be considerable slapping as small waves and wash from passing traffic bounce back off the underside of the hull when you're at anchor. It's worse still if you are on passage and motor-sailing, as the engine will be grinding away right next to your ear!

Cockpit

On smaller boats the cockpit will inevitably become part of the boat's accommodation, especially if you have a cockpit tent cover to protect it from the weather. I've slept quite comfortably on the cockpit benches of a 16-footer in the past, so a good tent with see-through panels is an excellent investment. Kids love it too – it gives them the feeling they are camping – although I'd be a little wary of letting them sleep there on their own when they're very young just in case they decide to go walkies!

In terms of both comfort and security it's good to have a cockpit with high coamings (seat backs) and a deep foot-well. Also, if you plan to take your boat out into the open sea it's important to ensure it has some means of stopping water from going down below (a fixed washboard or bridge deck), plus the cockpit must have some way to self-drain should a large wave come over the top and into the cockpit.

Multihulls

With two hulls in which to place bunks, heads, galleys, navigation stations etc., plus a bridge-deck saloon that is on the same level as the cockpit, most cruising catamarans have more living space than a monohull of equivalent length. When compared with earlier models, modern cats have wide, flat-bottomed hulls that can easily house double berths and large, luxurious heads. Furthermore, the almost full-width deck saloon of a modern cat has enough space to house a navigation station, a luxuriously equipped galley and seating around a table for six or eight people in comfort.

Deck space is also vast compared with a monohull – particularly in the cockpit and on the foredeck. A multihull's cockpit is very wide compared with that of a monohull and might possibly seat 10 or more people for socialising. Consequently there is more than enough stowage beneath the seating for all the usual cruising paraphernalia such as inflatable dinghies, outboards, buckets, mops, snorkels, barbeques etc. Often you'll see cruising monohulls with a lot of these bits and pieces tied down to the deck due to lack of locker space. With a few minor modifications such as some strategically placed netting, a multihull's cockpit is easy to turn into a safe play area for young children. There's often enough room to inflate a small paddling pool as well, so even the youngest nippers can have a splash around.

Catamarans have bags of room in their cockpits

The foredeck of a modern catamaran is also very likely to have an area of solid deck for sunbathing, anchor handling and stowage nowadays, as well as the usual netted trampolines (simply the best place to keep cool on a hot night in the Tropics!).

Trimarans

To be honest there are only a few cruising trimarans in existence, or being made new today, and there is a pretty good reason for it too. Whilst trimarans are often capable of pretty impressive speeds, when it comes to accommodation they are usually worse than a monohull of the same hull length. This is because the central hull – the only one that provides any living accommodation – is normally shallower than a monohull in order to keep drag down to a minimum.

A trimaran can offer exhilarating speeds and if they fold up small – even better

The other downside is their beam, which is even wider than that of the same length catamaran. That said, most non-racing models are designed with folding floats, so yachts such as the Dragonfly, Corsair, Farrier and Telstar ranges can be moored up in a normal-sized berth, which helps to keep marina costs down. The smaller ones can even be trail-sailed on purpose-built road trailers and are light enough to be towed by a large saloon car or 4×4.

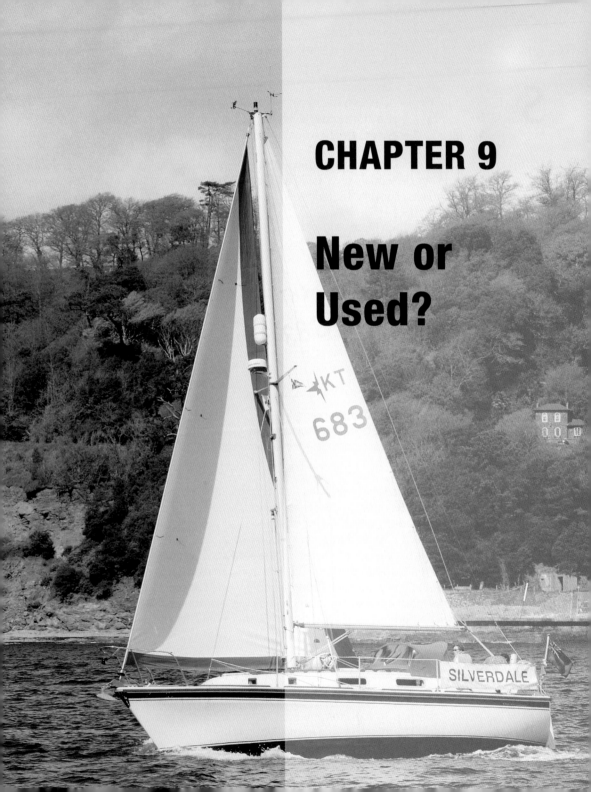

CHAPTER 9

New or Used?

S ome exceptional deals are available on new boats these days, so why buy a used boat? Well, although it's nice to own a new boat there are many good reasons for choosing a well cared-for used boat. For a start, most will have had any initial faults rectified and are likely to be already well equipped with safety, navigation and other luxury items.

The main drawback of a used boat is that you won't necessarily know how well she's been looked after. Depending on age, essentials such as rig and engines might be well worn and need replacing, as will pumps, toilets, cookers, batteries etc.

Privately owned boats less than three years old tend to be well shaken down, but not generally used to the point of requiring imminent major surgery. For this reason they usually make good buys, as do one or two year-old cars. If there were any warranty issues early on, these will have been sorted and proven by this stage, so you should be getting a problem-free, well-oiled boat ready for you to add your own personal touches.

Boats over 10 years old will quite likely be reaching the point where they need some major items replaced, such as sails and standing rigging, along with a bit of updating and sprucing up. Those that haven't been suitably updated might be available for less money, allowing you to choose exactly what you want in the way of new sails etc. So don't discount cheap, but slightly tatty boats, as sometimes you can benefit from the previous owner's frugality or lack of care.

Boats older than 15 years that have just undergone their first major refit and are sporting new sails, rig, engines etc., can also be a good buy, providing the hull has been maintained in good condition. Beware of those that still have most of their original equipment, though, as most of it will be nearing the end of its useful life.

Most marine equipment, especially engines, last a good deal longer if the boat is used on a regular basis, as there is little opportunity for anything mechanical to rust or seize. The exception to this rule is ex-charter boats, which might well have been used to the point of exhaustion. A typical ex-charter yacht will have endured ten times the wear and tear of a privately owned boat, although some of this is usually counteracted by the amount of regular maintenance needed to keep them fit for charter. Some of the larger charter companies completely refurbish their boats before they sell them, but they rarely replace major items such as masts and engines. Never buy an ex-charter yacht without a full survey, as there may well be some hidden GRP hull repairs that haven't exactly been carried out by professionals. Charter boats, like hire cars, don't necessarily get treated with a great deal of respect!

New-boat Budget

There is no point in starting to look for any boat until you have a good idea of what you can afford and what you are likely to get for your money. You will need to know how much you can spend

An ex-charter boat can be a good deal, but make sure she's been well maintained

on the boat, *including* the expense of equipping her with essential safety gear. When buying new it can typically cost at least another 15% of the purchase price to fully equip her. Many 'boat show bargains' come without crucial safety items such as flares, lifejackets, **liferaft**, horseshoe buoys etc., and are often light on other essentials such as instruments, anchors, chain, mooring warps, fenders and VHF radio.

The more desirable 'luxury' items such as a sprayhood, hot water, fridge, heating, cockpit table etc., will almost always be on the options list, and one of the most expensive additions will be electronic instrumentation – chart plotter, radar, Navtex, autopilot etc. Some racing yachts might not even include sails, as often the dealer will say that their customers prefer to choose their own. This might well be the case, but if you're after a performance boat, do make sure you've budgeted for this before you pay your deposit.

When buying at a boat show the price displayed will often be for the barest minimum, unless the deal includes a boat-show 'extras pack'. Prices invariably exclude delivery and commissioning costs, which can be quite a hefty extra sum. The average cost of delivery from mainland Europe to the UK is around £4000–£5000, and commissioning can add a further £3000–£4000.

Ordering your extras with the boat could give you a much better deal

So the best way to set a budget for buying a new boat, is to work out what you can afford in total, then reduce the figure by 20% for extras. Visit all the boats on offer that interest you within that figure, collect the standard inventory and options lists, then sit down somewhere quiet and add up the right-hand column once you've ticked off your preferred options. If there is still something left in your budget pot then you can go back to the top and tick off any 'luxury' items in order of preference until the pot is empty. Don't, however, dismiss the idea of opting for a slightly smaller boat with a more comprehensive inventory. If you buy the biggest boat you can afford with the intention of adding the goodies later, it will undoubtedly cost considerably more than having them fitted at the boatyard or while she is being commissioned – and particularly if there is a generous offer on yard-fitted extras at a boat show.

Typical extras required on a new boat

Essential/safety items (Approximate cost £4 000)
Horseshoe lifebuoy × 2
Liferaft (Can be hired)

Dan buoy
Flares
Lifejackets and harnesses
Sharp knife
Spotlight and 12 V cockpit socket
Radar reflector
Signal cones and balls
Kedge anchor and chain
Warps × 4
Fenders × 6
Dinghy, oars, outboard
Powerful torch
Automatic electric bilge-pump

There will be a lot of extra items to buy once you take delivery

Fire extinguishers
First-aid kit
VHF – fixed and handheld
GPS – fixed and handheld
Charts, pilot guides and almanac
Charting equipment (e.g. dividers, parallel rules etc.)
Storm sails
Bolt croppers

Luxury items (Approximate total cost £25 000)
Hot water
Sprayhood
Bimini
Additional battery capacity

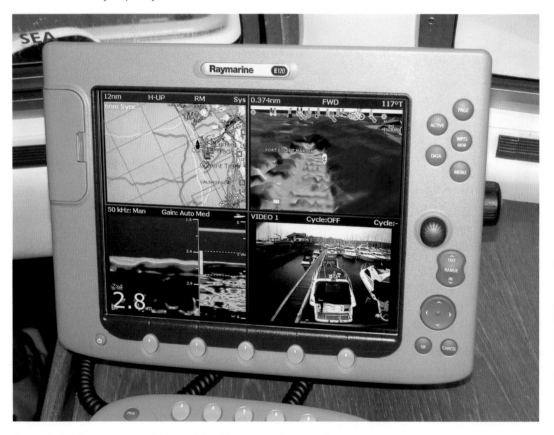

Some desirable extras can add a considerable sum to the overall cost

Chart plotter
Navtex
Radar
Automatic Identification System (AIS)
Teak decks
In-mast furling
Electric windlass
Autopilot
Cockpit table
Transom shower
Water maker
Air conditioning
Shore power
Television
Spinnaker/cruising chute

Used-boat Budget

To some of you this may seem like teaching Grandma to suck eggs, but in my experience it pays to do your homework before you start wandering around the boatyards and climbing over the boats for sale – however tempting it may be.

Before you even contemplate looking for a boat, set yourself a rigid budget that you can afford. It might sound a little obvious, but I've lost count of the number of times I've seen people spend so much on the boat itself that they can't afford to equip or run it properly when they finally take possession. Remember, the bigger the boat, the higher the running costs, so do be realistic.

When looking to buy a used boat it's a good plan to start with some idea of the value of the type of boat you want – in its most basic form. This you can usually derive from looking at a wide selection of similar boats for sale. It's likely that the boats priced at the lower end of the scale have less, older, or more worn equipment, so the price will more accurately reflect the true value of the boat, rather than its inventory. Mentally reduce any asking price by your estimated value of the inventory second-hand, then knock off a further 10% to allow for the inevitable bargaining and you should now have arrived at the approximate *Base Price* of that model of boat.

If you are looking at a popular boat of which many were built and over a long period of time, then it might pay to split these estimates into age bands, or even by the exact year (particularly if the model was updated frequently), to arrive at a more accurate figure.

Now, for each boat you see for sale you will need to work out the total value of all the extra equipment being offered with the boat and add it back on to your previously calculated *Base*

Model	Name	Launched	Asking Price	Base Price	Inventory	Offer Price	Extras	Overall Cost
W Centaur	Freedom	1975	£11995.00	£9500.00	£1500.00	£11000.00	£1500.00	£12500.00
Sabre 26	Sarah	1977	£10495.00	£8000.00	£1000.00	£9000.00	£2000.00	£11000.00

Creating a table of costs will help you remember true values when deciding

Price, to arrive at your *Offer Price*. This will obviously vary from boat to boat, depending on its age and the quantity and condition of any ancillary gear.

Work out if the boat can be brought up to the specification you require, within your overall budget. Make a table showing the cost, if any, of replacing essential and expensive items such as the engine, sails, batteries, dinghy, outboards etc., and add these to your *Offer Price* to give you an *Overall Cost*.

Create a simple table showing any extra equipment or repair costs and a column showing the 'actual' cost to you of each boat.

Where is Best to Buy?

Nowadays the Internet has made life easy for those looking to buy a boat, car or house. There's a fair chance that over 80% of boats up for sale will be on the web somewhere, so it's the obvious place to start looking.

Just searching for a boat model doesn't always work, however, so you might need to juggle your search terms. When you discover a web page showing an interesting boat for sale, follow any link to the broker if possible, as this often opens up another great source of boats and other models you might not yet have thought about, or even heard of.

Once you've found a few boats you like the sound of, search for an owner's association for that model, as these can be a veritable gold mine of valuable information that no private vendor or broker is going to mention. Often they list all the likely problems associated with a particular boat, or a particular model of that boat, telling you whether it can be fixed and for how much.

All this is invaluable to you when you first set foot on the boat, as you'll be able to have a subtle look around the area in question, to see if there are any obvious signs of repairs having been carried out. If there are, you can ask the vendor directly what has happened and why. If there aren't any obvious indications, you will still want to add it to a list of pointers for the surveyor, should you be interested enough in the boat to appoint one.

Some knocks can be fatal, but others can be worth repairing at the right price

Don't be scared off by signs of repair work having been carried out until a surveyor has seen it and decided whether it has been done properly or not. If an experienced shipwright has done the job there's a good chance the area of repair could be better and stronger than it was originally.

Typical problems to look out for are repairs around deck-gear fixing points. If the sealant has decayed then water can seep into the balsawood sandwich filler commonly found in yacht decks to make them lighter and better insulated. If the area has had all the rotten balsa or plywood backing removed and has been properly fixed using new wood and epoxy, then it's a problem that's unlikely to recur for a long while. Similarly, examine the mast support. Quite often the area of coachroof around the foot of the mast becomes depressed from the rig loads after a time, especially if she has been raced a lot and had her rigging 'tweaked' a little over-enthusiastically. On some boats this is a common fault, but can easily be remedied by fixing extra support in the form of a crossbeam or upright directly beneath the mast step.

Sometimes small deformities can look worse than they are, so try to be constructive about your investigations. Too many first-time buyers are scared away by dripping windows, dodgy stanchions, leaky stern glands and the like – all of which are easily fixed, but give you some useful bargaining power when making an offer. In the past, when my budget has been severely restricted, I've deliberately looked for boats that appear to be in a state of extensive disrepair. The mouldier the better was my motto! This is because the owner has either given up sailing or bought another boat and can't cope with the cost of maintaining both. This is a good sign that the vendor is pretty desperate to accept any reasonable offer.

Value Retention

Another point worth bearing in mind when buying a boat is its inherent resale value. Top-quality boats might be the most expensive, but then they are usually the most sought after on the second-hand market. In fact a number of boats that are known for their build quality, such as many of the Swedish cruising yachts – Halberg-Rassy, Najad, Malo, Arcona – and some UK-built, semi-custom boats – including Rustler, Southerly and Oyster – are some of the quickest to be snapped up as it is well known that they hold their value. The point is people trust and aspire to these yachts, so their value rarely drops on the market and you are more likely to get a good price back when you sell her sometime in the future.

Equipment, however, doesn't always increase a yacht's value when selling. A vendor might have bought a brand-new Southerly and installed the very latest electronic navigation system at the time, but five years later most of the gizmos will be old hat and already looking dated, so don't expect to get your money back for extras unless they are non-electronic items such as electric winches and anchor windlass, dinghies, liferafts or similar.

In fact it's quite common for owners to take some of the more expensive portable items with them if they're buying another boat, so you'll need to study the inventory carefully. On this point – unless it is stated otherwise, when you go on board a boat that is advertised for sale, any items on board that are mentioned in the sales brochure or inventory become legally part of the sale contract. When you make an offer you are doing so for the yacht and her entire inventory of equipment as listed, so if the vendor is disgruntled and thinks your offer to be too low, but he grudgingly accepts it because, maybe, his mooring contract is about to expire, he is not allowed to remove or exchange any items that were on the inventory at the time you made your offer. If any equipment is missing when you take possession of the yacht, then you are entitled to demand either financial recompense from the vendor or a replacement item of the equivalent type and value.

More often than not a better quality yacht will hold its value for much longer

I heard of a case where the vendor had to accept an offer he didn't like, so he removed all the spare sails from the boat before he handed it over. Having sold these to other members of his club, he was then forced to buy them back to give to the new owner, or be sued for misrepresentation.

CHAPTER 10

Additional Costs

Brokers and Agents

A good broker, preferably one listed with the Yacht Brokers, Designers and Surveyors Association (YBDSA; www.ybdsa.co.uk), can help considerably with your search and will know all the relevant paperwork that needs to be completed before, during and after a purchase. Brokers can help when it comes to establishing if there are any outstanding liens (mortgages), legal encumbrances, debts or writs that the boat might have, and they should also be able to help with registration and possibly even with marine financial advice.

The downside is that you might end up paying more for the boat because the vendor will be paying the broker a fee in the region of 5–8% of the sale price and will most likely be looking to

Going through a broker can save a lot of hassle

recoup some or all of this amount in the deal. Also, with a brokerage deal you can often miss out on a one-to-one chat with the previous owner, who might well have given you some very helpful advice about the boat. That said, you at least have a legal comeback against a broker if the boat is not of merchantable quality, which you won't have if you buy it through a private deal.

Surveyors

To instruct a surveyor when buying a boat is not mandatory, but is highly advisable on a used boat of substantial value – after all, you wouldn't dream of buying a house without one. Some also advise using a surveyor when buying a new boat, but this isn't common practice unless you're paying top dollar for a custom-built yacht.

If in doubt as to who to choose or where to find the right one for the type of yacht you're looking for, ask around in yards, clubs and boatyards for a recommendation. Also, check with the YBDSA as a listing there is the most reliable proof of a surveyor's competence.

Qualified surveyors will quite likely have letters after their name as well, through which they can easily be checked – e.g. MRINA (Member of the Royal Institute of Naval Architects, www.rina.org.uk), or MImarE (Member of the Institute of Marine Engineers, www.imarest.org).

If you are buying a specialist boat, it is worth finding a surveyor who has experience of that category of vessel, particularly with wooden or metal boats, and also with some multihulls.

A marine surveyor will offer a choice of surveys depending on the depth of search you want – and the charges will reflect this. The least expensive is just an overall condition survey, in which the structure of the boat and its seaworthiness is checked. It does not cover anything that is inaccessible without removing parts of the boat and cannot include an inboard engine if the boat is out of the water. If in doubt about rigging or engines, commission a separate survey on these – particularly if a sailing rig is over 10 years old – as you might need this for insurance purposes. If in doubt about anything in the report, ring the surveyor and discuss it with him in detail.

Do get a qualified surveyor to take a look before you agree to a deal

Once you've revealed the boat's true condition you can start to bargain with the vendor if there are faults that need rectifying. If there are serious problems such as structural damage, osmosis, insecure rig or a faulty engine, don't necessarily be put off buying her – try getting quotes on the necessary repairs and reduce your offer accordingly.

Financing your purchase

Most brokers have their preferred finance companies and can arrange the whole thing for you. Their quote might well sound reasonable, but it's worth ringing around for comparisons before committing yourself. Unsecured loans up to £25 000 with a fixed or variable interest rate are available from a variety of outlets, so shop around. No survey or registration is necessary with unsecured loans, but they might be more difficult to acquire without verifiable proof of your income and maybe a substantial deposit.

For loans of over £25 000 you will have to take out a marine mortgage or secured loan, although there's nothing to stop you having a mortgage for less than £25 000, particularly as the interest rate will be lower on a secured mortgage than on an unsecured loan.

Marine finance brokers will have a greater understanding of boat buying than an ordinary bank or loan company, so they can advise you on the best way of going about the transaction. As with a house mortgage, a marine mortgage will always be secured on the boat and, because of depreciation, you will rarely be granted a loan for more than 80% of the cost.

There are a variety of repayment methods available. Capital repayment schemes can have 'equalised' repayments that remain constant regardless of interest-rate variations. If the interest rate drops, instead of the monthly payment reducing, the period of the loan reduces – and vice-versa.

There are low-start mortgages with a discounted interest rate for the first few years, which used to be ideal for those planning to repay the mortgage early. These days, however, there are often penalties for early repayment so do check the amount of any early redemption fees you might be liable for before signing.

A 'balloon' mortgage can be obtained whereby you pay a deposit, followed by low monthly repayments, with a lump sum (usually 40%) left to repay at the end, before you become the official owner. This is useful if you are expecting a windfall at some point in the future, but will prove more expensive over more than three years.

As with house loans, some finance companies allow payment holidays and lump-sum repayments during the life of the loan, although the amount you can repay in one sum is usually limited and there is a fixed period before you can make the first payment. If this is what you intend, discuss it in detail before signing up to it and check the small print carefully for any penalties that might be payable.

Boats for which a mortgage is required have to have full Part 1 registration – an SSR (Small Ship's Register) number will not do. This isn't difficult with a brand-new boat, because it can be done easily from the start, but it can be more of a problem with a used boat as you'll need to prove the boat is clear of any liens before you can Part 1 register her. Most marine finance houses can check the boat is free from any debts on your behalf, however, before any agreement is signed.

Boats over a year old will also need to undergo a survey and professional valuation.

Mooring Types and Costs

Probably the boat owner's biggest continuous outlay will be mooring fees, unless the boat is small enough to park up on a trailer in a garden or drive. Certain types of moorings (particularly on the UK south coast) are becoming scarce, so costs are often fairly exorbitant. Depending on the size of boat you plan to buy, mooring options for non-trailerable boats are:

- a tidal or non-tidal swinging (buoy) mooring
- a pile mooring
- a pontoon berth between piles
- a marina pontoon with all facilities

Piles and buoys

Charges increase according to convenience, so if you're happy to leave your boat swinging on a buoy in a river or harbour and row out to it with all your gear every time you use it, this will probably be the cheapest solution. It can be even cheaper if you join a yacht club and are allocated a mooring – although waiting lists for moorings are often long, even if you're a member.

If your boat can safely take the ground and you don't mind being slightly restricted in your sailing times, then a half-tide mooring (accessible, say, from 3 hrs before to 3 hrs after High Water) is even less expensive.

A swinging mooring can be delightful and save considerable fees

Another benefit of a swinging mooring is that you can sit on board eating a meal, or simply taking in the view, without even leaving the harbour.

The downsides, though, are:

- Lack of security – more gear/boats get stolen from remote swinging moorings than any other.
- Remoteness – you will have to keep a tender in a nearby marina or yacht club, or park your car close by and pump up an inflatable dinghy each time. Either way you will have to row/motor out to the boat with gear, provisions, tools, and crew etc., sometimes making several trips. Furthermore, you will then have to take the dinghy on board, tow it behind or leave it on your mooring while you're out sailing and take a chance on it disappearing.

Moorings between piles are often more secure than buoys

- No power or water – to keep your batteries charged you will need either a wind generator or solar panel and you'll need to fill your water tanks at the fuel pontoon or carry water on board in jerry cans.
- You have to ensure the ground tackle and the boat's own mooring strop is regularly inspected for wear, otherwise you will never sleep on windy nights – whether you're on board or not!
- Your insurance might well be higher, especially outside of the usual boating 'season' (1 May–30 September).

Pile moorings have the same pros and cons as buoys, although the physical security is slightly better in that the boat will be attached to the ground more substantially and from two different points. Wear on the strops, though, can be equally ferocious in bad weather and often you will be rafted up alongside another yacht, so your fenders will take a hammering.

A pontoon suspended between piles makes boarding and disembarking with gear and crew considerably easier. It is also easier to carry out work on the boat and to clean it. However, the wear on fenders and lines can often be greater than swinging or pile moorings. Remember, also, that you will most likely want to store your boat ashore for part of the winter at least, so craneage and hardstanding storage charges must be added to your annual moorings budget.

Marinas

Boat owners who have a busy working life usually prefer a marina berth as they can arrive and jump straight on board. The batteries will also be kept charged using shore power and someone from the marina will walk around and check your lines and fenders regularly – particularly in strong winds.

You can wallow in the marina showers and take a hot meal ashore while your gear is being washed and dried in the launderette. It is simple to have work done on her whilst you are away from the boat and if you are doing it yourself there's usually a chandler on site or nearby for bits and pieces. All this luxury comes at a price, however!

In addition to the cost of the berth, other marina costs can include storage and electricity. Some marinas include a certain amount of time ashore in the annual fee, while others only offer a discount off the cost of dry storage to annual berth holders. So if you want to keep the boat out of the water from October to April, check the charges with the marina first and don't forget there will be haul-out and re-launch charges as well, which are often quite high in many of the smarter marinas.

Sometimes electricity is included in the annual berthing charge (usually only up to a limited amount and at a restricted capacity), but more often an additional charge is made. Furthermore, you might be charged rental on the metered lead, which only the marina can supply.

Many find the lure of the marina just too tempting!

Winter Maintenance

If you want your boat to remain in tip-top condition you will need to maintain her prior to, and during, the winter months. You will need to have the engine fully serviced and the oil changed before winterising her, otherwise acidic residues in the old oil can damage the engine.

Either disconnect the batteries and take them home to keep charged over the winter, or set up a proper marine battery charger that has a maintenance programme, as batteries self-discharge considerably quicker during the long, cold nights. If you allow the batteries to become fully discharged, it's likely you'll be forking out for new ones in the spring.

Taking off your sails and sending them to be laundered and stored each winter will probably add years onto their life and save you money in the long run. The same with any canvas items such as sprayhoods, sail covers etc. A sprayhood left up all winter will degrade five times quicker than one that's cleaned and stowed in the dry – and at around £800–£1 000 a time, it has to make sense!

Budget for more financial outlay in the spring, when it comes to cleaning, polishing, antifouling, replacing anodes, servicing seacocks, replacing hoses and hose clips, servicing lifejackets and the liferaft and replacing any worn running rigging. Roughly speaking each spring fit-out can cost between £200 and £500 for a 25-footer, £500 and £1 000 for a 35-footer, and so on. But if you're careful and look after all the ancillaries it's surprising just how much you can save over the years.

You won't be alone in the spring, when the bulk of the maintenance needs doing

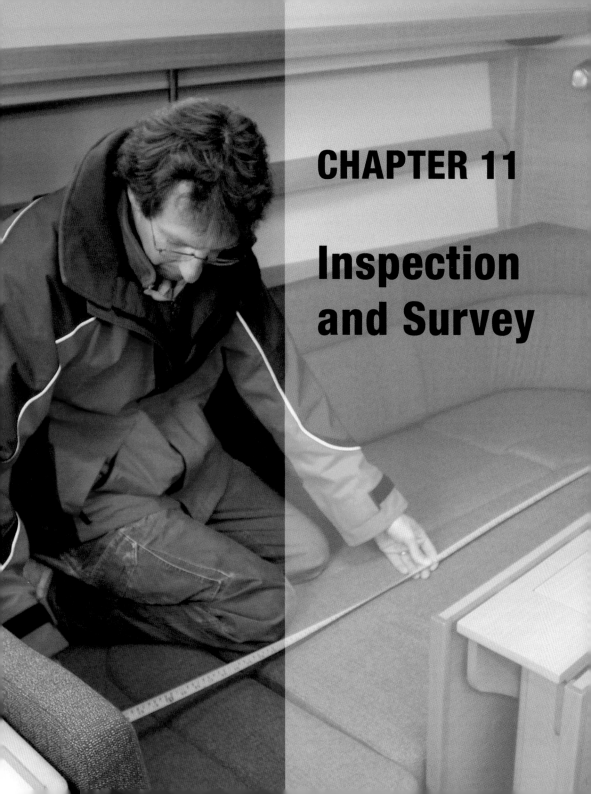

CHAPTER 11

Inspection and Survey

A huge amount of money can be saved during the boat-buying process by carrying out a competent and thorough initial inspection yourself. Some vendors can get a little tetchy when you really start to open things up to get at the nitty-gritty, but if they do I'd be a little concerned that they had something to hide and might, at that point, be tempted to just walk away.

Some years ago I took a Diploma in small craft surveying at Plymouth University, which added further knowledge and authority to my boat testing abilities. The course, while a little pricey for someone who just wants to get an overall idea of how good a boat is before instructing a professional surveyor, is a valuable tool for those who regularly change their boats and could even save you a fortune in failed professional surveys. You might even want to put those new skills into practice one day by becoming a professional surveyor yourself!

Self-surveying

The initial dilemma with inspecting any boat is that it'll either be in the water or out. Obvious you may say, but if it's in the water you won't be able to check the underwater part of the hull, and if she's out – well you've no idea whether she leaks like a sieve or if the engine will run at all.

Personally I'd rather see a boat out of the water first, as it's more important to be able to inspect the hull, keel, rudder, skeg, prop and anode/s than know if the engine is good or not. Engines can be changed relatively easily, but serious flaws in the hull are a great deal trickier to remedy.

The hull

So, starting with the hull, just stand directly in front of her and look down the centreline of the hull. Does everything look okay? Is the hull symmetrical and are the keel/s, skeg (if she has one) and rudder in alignment? I went to see a small cruiser once and the owner was immediately keen to get me on board – holding my arm and nudging me towards the ladder. I declined, saying I just wanted to take a few photographs of the hull first, so I'd remember what she looked like after I went back home. Standing about 10 m in front of the boat (a 24 ft GRP bilge-keeler) I could clearly see that one keel was more vertical than the other. On closer inspection it was clear that some GRP work had been carried out around one of the keel/hull joints. In fact the new GRP actually overlapped the keel joint by an inch or two – which immediately had me concerned. I went straight on board and lifted the inspection panel in the saloon **sole** – to find a large area of very poorly applied GRP all around the keel root area. In fact the job had been carried out so badly (the laminator hadn't even bothered to clean and dry the area of old GRP before applying the new) that the edges of the laminate repair were lifting off the hull.

It was abundantly clear to me that the keel had taken a big wallop – possibly by being dropped onto the hard – which had badly damaged the hull around the keel root. Not the end of the world had it been repaired properly, but the poor repair job and the vendor's dishonesty soon had me high-tailing out of the yard!

I don't mind if a yacht has been damaged at some time, providing (a) the owner tells me – and preferably before I've driven 50 miles to see the boat; (b) he shows me some paperwork that at least indicates it has been repaired properly; and (c) he lets me take a very close look at it. Oh, and (d) he adjusts the asking price accordingly!

The first task on a GRP boat is to study the hull for dents, gouges, new patches of gelcoat and, probably most importantly, blisters or evidence of repaired blisters. It's quite a tricky job to disguise little dabs of new gelcoat that have been applied where a blister has been opened up and repaired, so they should be easy to spot. Beware the newly painted hull, though. If an owner

Sure signs of advanced osmosis – these blisters are a giveaway!

isn't willing for you to scrape it away over a suspicious area, then I'd start to worry. No-one who is seriously selling his boat will put fresh paint on her hull, knowing full well that a surveyor will need to get beneath it for a proper hull inspection. Anyone who does is either very naïve or hiding something.

Clearly you can't gouge great lumps out of his gelcoat, but a little judicious scraping here and there can reveal an awful lot. This is another good reason for getting the surveyor's diploma – people are a lot happier if you can prove you know what you're doing.

So, after inspecting the GRP for any signs of damage, repairs or osmosis, you should look at the leading edge of the keel, towards the foot, for any signs of grounding or damage caused by heavy floating debris. Also, check the keel joint for seeping rusty water stains. These might indicate that perhaps the keel/hull seal has gone and/or the keel bolts need replacing, but this isn't always the case, so don't be unduly frightened by a few signs of rust. Most externally attached fin keels are made from cast-iron and they rust like nobody's business. It is a never-ending job to try to keep them from oxidising and few will look good when they're lifted out at the end of the season. However, do ask the owner if he has had a keel bolt extracted recently for closer inspection and remember to check the tops of the keel bolts from inside in the bilges when you go aboard. A rusty bolt head is often a sign that they'll all need replacing soon.

Take a close look at the anode, which is a large lump of zinc bolted somewhere on the hull below the waterline, or possibly an oval-shaped one fitted directly onto the prop shaft. If it looks exhausted – i.e. there is little zinc left on the mounting – I would look even more closely at the propeller and skin fittings (water outlets or inlets in the hull for sinks, loos, engine cooling etc.). A worn anode can mean there has been some galvanic corrosion present, which might well have allowed metal items such as bronze skin fittings and propeller to 'dezincify' and/or corrode. The zinc anode is sacrificial – i.e. it is deliberately there to absorb any corrosive action caused by different metals reacting together in the saltwater. Note that an anode should be connected electrically to all the metal items that it is protecting, or it won't work, so make a note to check for this later when you're on board.

If any of the bronze skin fittings have turned a rosy pink colour, outside or inside, it is a definite sign of dezincification. This is where the zinc in the alloy has dissolved into the surrounding seawater through lack of anodic protection. Beware, this can be extremely dangerous in that the fitting itself might break off inside causing a serious leak and possible sinking of the vessel.

Check closely around the rudder area and 'waggle' the rudder to check for excessive play in the stock bearings or pintles. There will quite often be a little play, but more than one millimetre and I'd jot it down to point out to the surveyor, should I be interested enough in the boat to instruct one. Do the same with the propeller if she is shaft-driven and pay attention to the bearing in the P-bracket (called a cutless bearing, it is the one through which the propeller shaft goes, just ahead of the propeller itself), as this is a common point of wear.

The anode protects the metalwork beneath your boat so make sure it's good

A large majority of modern production yachts have a Saildrive power-unit fitted, which is where the prop shaft is inside a drive leg – rather like that of an outboard engine – and the whole leg protrudes from the hull through a series of seals. Close inspection of the seals is important, as these need replacing every five years or so and can cause serious leaks if they are left to degenerate.

With wooden hulls most of the likely problem areas are similar, but it's important to inspect the framework (scantlings as they are known) from the inside as well as the planking on the outside. Usually a wooden boat is more open inside than a GRP one, so it should be fairly easy to check the condition of the frames, deck beams etc. A common problem to look out for is split frames (ribs that hold the planking in place) where they have come under undue stress. Often these will have been doubled up, or reinforced around the bilge area, particularly on long-keeled yachts. It won't necessarily be disastrous if a few are cracked, but you should take it as an indication that (a) she might have been put under serious stress at some point and (b) the owner might not have maintained her all that well.

One other important thing to look out for on a wooden boat is the condition of the fastenings that hold the planks to the frames. Depending on how old and what style the boat is, most of the fastenings will need to be replaced at intervals of 25 years or so. First signs are usually a darkening of the plank around the rivet or nail, which, if left for too long, can damage the planking as well. I would definitely want a wooden boat specialist to look her over if I saw such signs. If the caulking is falling out between the planks of a carvel hull, it's not so much of a problem. Caulking is quite a satisfying task and one that doesn't require a massive amount of skill or experience – just patience!

Other key points to check on a wooden hull are where the transom joins the hull and the garboard strake (the planks immediately above the keel). Both these areas suffer considerable stress and are always prone to leaks and rot.

Recaulking the planks is a fairly frequent task with wooden boats

Decks

On wooden boats the prime problem is obviously rot, so you'll want to look closely at all the places where water is likely to gather or seep, such as hull/deck and deck/coachroof joints, deck fittings and any upstanding pieces of plywood such as cockpit coamings etc.

Look out for any areas of wood that have been GRP or canvas sheathed – particularly the side decks. If they feel spongy to walk on there's a fair chance the plywood beneath is rotten or delaminating due to water ingress.

On GRP boats you'll need to look out for stress cracks around deck fittings and chain plates. These are an indication that normal loads have been exceeded or the backing plates behind or within the structure have broken or rotted.

GRP decks on post-1970s yachts are almost universally made from a sandwich composite – i.e. two layers of GRP with a filler such as balsawood or high-density foam. The most common

Non-slip deck coverings such as Treadmaster can hide a multitude of sins!

problem with these is that the seal breaks around deck fittings, letting water into the sandwich and saturating the filler. This then turns into a mouldy 'mush' and the inherent strength of the composite deteriorates. While this can be cured – usually by stripping off one of the layers of GRP, top or bottom, replacing the filling and relaminating over it – it can be both messy and expensive. Sometimes repairs are done from on deck and a non-slip deck covering such as Treadmaster is stuck over the repair so as to make it invisible. So, if the boat you're inspecting has this type of covering everywhere, be sure to test for sponginess under foot and any signs of different-colour patches of GRP around the edges. If you're concerned, then the only way is to check the decks with a moisture meter, so jot it down to ask the surveyor.

The rig

Almost any used boat over 10 years old will need to have her standing rigging replaced to

satisfy the insurance companies, so if she is over that age do ask the owner or broker if there is any written evidence that this has been done. It doesn't necessarily mean that the rig is faulty or dangerous (I've sailed 20-year-old rigs after closely inspecting every inch of them), but the insurance companies more often than not stipulate it as a necessary precaution. This is because modern stainless steel rigging wire rarely shows any sign of imminent failure, unlike the older galvanised wire rigging, which rusted or had individual wires fray around the terminals.

If the rig is between 10 and 15 years old some insurance companies will accept a satisfactory electronic rig inspection (where the resistance and electrical conductivity is measured at each terminal), but not all.

If the rig is less than 10 years old, but not new, you will probably want to take a good look at it for any signs of wear or corrosion – although, as I say, it is often very difficult to spot where the wires or terminals might be weakening. If in

A specialist rig surveyor might be worth it – or you could do it yourself

doubt – and the absolute integrity of the rig is probably second only to that of the hull – I would get it electronically checked by a professional. Check to see if there is a surveyor nearby who has the skills and equipment to do it as well as the usual hull survey, as it is quite likely to be less expensive than instructing two separate surveyors.

You can check it yourself at deck level, however, so it's worth taking a look at all the typical stress points – chain plates, deck eyes, gooseneck (where the boom attaches to the mast), kicker mounts, etc. It might also pay to take a glance at the rest of the rig through a pair of binoculars as well, just to see if you can see anything obviously awry. Then play a little with the running rigging to make sure furlers, winches, rope clutches, jammers etc. are all fully functional.

Sails

The owner should be willing to let you see and inspect all of the boat's sails, out of their bags. After studying the seams, clews, chafe patches etc. of the working sails, make sure you look closely at the rarely used sails as well. If at all possible and it's not too windy, ask the owner to hoist/unfurl the sails if they are already bent on, so you can see them properly.

Often sails like storm jibs, trysails and spinnakers are rarely, if ever, used, so even the owner may not know their true condition. I pulled a storm jib out of its bag once on a boat I was thinking of buying and it nigh-on disintegrated. Not only had it been put away wet, it was screwed up with several metal attachments inside the sail. The sail was almost entirely black with mould, except for the areas where a corroding **shackle** or block had changed it to rusty red. The clew ring actually fell out of the sail when I pulled at it to get it out of the bag!

It's always good to see bills for new sails, or even cleaning and winterising, to boost your confidence that they have been looked after properly and aren't likely to fall apart the day after you've parted with your money.

Below decks

The first thing to look at is the bilges – are they dry, and if so are there any signs of water ingress such as dirty stains? Does it have an automatic and a manual bilge pump? Check the keel-bolt heads for rust, although some will be laminated over so it might be difficult to tell. In a wooden boat you'll need to see if there are any little corners where water has been unable to drain and has penetrated the wood. Unfortunately these will always be the areas most difficult to access. Telltale signs include bunged-up limber holes where the water has been unable to drain into the bilge.

Look closely at all the hull skin fittings and seacocks etc., checking for dezincification, corrosion and, if the boat has cheap steel ball-type seacocks, rust. All through-hull skin fittings should be of a proper marine grade, preferably made from bronze, with bronze seacocks to match. Gate

There might be no hope for some old sails like this one

valves and cheap plumbing stopcocks are not suitable for boats and could endanger the vessel and the crew if they fail catastrophically at sea.

While you're looking at the skin fittings, check to see if they are electrically connected to the sacrificial anode, as ideally they should be. If not, they could be corroded inside.

On most GRP boats the bulkheads are physically bonded with glassfibre onto the hull, so check all around the bulkhead to make sure these are still firmly attached and there are no signs of it coming away. If there were, then I would look all around that area for a possible cause, rather than just being concerned about it parting company with the hull or deck. In some cases the rig might have been over-tightened and this has put a strain on the mast step, which in turn will put extra load on the bulkhead, as well as cracking the area around the mast step.

After the bilges, check around the water and fuel tanks for corrosion and broken mountings, as to replace one of these can often be a major task if they have been installed prior to the decks going on. Sometimes it can even result in having to cut a large hole in the deck to remove the old tank, but this is rare.

Other things to look out for are the headlinings – these tend to part company with the deck-head after 10–15 years due to the glue breaking down or the foam backing disintegrating. It's not necessarily bad news, but it's a tedious job to replace it all, especially if half the remaining old stuff doesn't want to come off. I had one boat in the past on which the owner had epoxied cork tiles everywhere, thinking it would insulate the boat better in the winter. Of course after a few years they started to peel off in places and ended up being a nightmare job to get off and start again. A boat with its headlinings hanging down can look pretty rough and put a lot of people off buying her, but to me it's often a good thing as the owner is more likely to accept a reasonable offer.

Leaking windows and deck fittings are some of the most annoying and possibly destructive faults with older boats. Whenever I find a leak in my boat, which is fairly frequently I might add, I do my best to stop it immediately, before water gets the chance to seep in behind furnishings or into lockers etc. and stain or spoil much of the interior. Otherwise it'll always end up being a much bigger job.

If you're looking at a boat and the curtains are stained, or there are water-marks inside the lockers, this is a pretty good sign that something has been leaking – and for how long? As I said earlier, leaking deck fittings can cause the sandwich deck fillings to absorb water and rot, so if they're not sorted quickly you could have a serious mushy deck problem to contend with.

Wiring and electrics in general always cause problems in boats – even new ones. Nowadays most boatbuilders are a little more careful about how they wire a boat, but not all. I visited a yard a few years ago to find they were using vacuum infusion techniques for making their decks. A great idea for saving space and weight, I agree, but then they were laying all the lighting cables

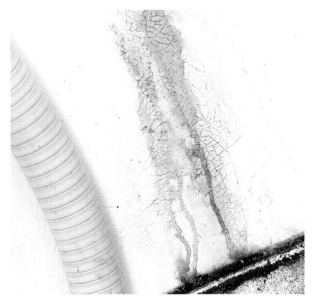

Use the evidence of leaks to track down any possible damage

in the bag as well and sucking them into the moulding at the same time! The result – absolutely no chance of pulling out a faulty wire to replace it, and worse still they hadn't even put in a few spares. Luckily these days most builders bond in trunking to allow cables to be replaced or new ones added. Whatever the age of the boat you're looking at, do take a look at the wiring where you can see it, and if the switch panel is easily hinged open, take a look behind to see if it is neat and tidy or just a box of spaghetti. Rewiring a boat is a tricky and expensive job, so it's worth finding out as much as you can about the existing system from the owner – and I would certainly want to see some notes on any non-standard additions to the circuit on yachts younger than 10 years.

While you're at it – take a glance at the date stamps on the batteries. Leisure batteries usually last around 4–5 years, depending on how hard they've been worked and how well they've been charged. If they're close to their 'duff by' date you'll need to add it to the extra costs list.

Ergonomics

An important part of a cruising yacht is whether you can live with it, and more importantly in it. It doesn't matter so much on a day-boat or a weekender, but if you're planning on living aboard for extended periods you'll need to consider certain aspects of the yacht before you leap in with an offer.

Is there enough headroom in the crucial places – galley, heads etc. – or will you be cursing the 'sticky-out' bits every five minutes? Can you get dressed in your cabin or will you have to do it in the saloon? Can you lie fully outstretched on the bunk and could you get out of bed without climbing over and waking your partner? All these things can make living on a cruising yacht a nightmare after just a few days, so walk around below, try to simulate living on board and get a feel for what being on board for several days, day and night, will be like. If you're looking to buy a new or nearly new yacht, and you're not sure whether it will be suitable, why not charter or crew on one if you can to see if it really is as good as it looks? A few days and nights living on board in real conditions will tell you a whole lot more than 10 minutes walking over her in a boatyard.

One of the most frustrating things we find, as a family living on board for months at a time, is a lack of organised stowage. Big gaping bins under the bunks and settees might be great for large items such as inflatables, spare sails, jerry cans, toolboxes etc., but they're useless for small bits and pieces such as rig spares, blocks, shackles, torch batteries and the like. We solved the problem by buying loads of different-sized plastic boxes with good lids and worked out exactly where each would fit and what it would contain. Then (I know it might seem fairly banal and unimportant to some) we labelled each box and stowage bin so we could find anything we needed pretty quickly and with the minimum of unpacking/repacking. This was also a perfect solution for storing clothes. Sealed plastic boxes are cheap and excellent for keeping the contents dry and sweet, even if they're stowed in the bilges.

Stowing your gear in boxes keeps things organised and dry

Be prepared

The more thinking you do before embarking on the hunt for the yacht of your dreams the better. These days we have the luxury of the Internet to help us, which is fantastic for finding out about a particular yacht and its foibles. This should enable you to get a pretty good idea of what a certain brand of boat looks and feels like without you having to kick around the yards traipsing over loads of unsuitable craft before you narrow your choice down to what you're really interested in.

Get yourself a notebook dedicated to buying your boat only and jot down everything relevant to your search – starting with your budget and finishing with the level of inventory you'd most like to see on board.

List the pros and cons of each boat you look at and discuss them at length with your family, partner, syndicate members and anyone else who is likely to be involved in your sailing. Also, don't forget to speak to other owners about their likes and dislikes before you make a final decision.

Equipment to take for an inspection

Camera
Binoculars
Notebook and pen
Small hammer
Wooden mallet
Paint scraper
Large and small screwdrivers
Pliers
Adjustable spanner
Multimeter

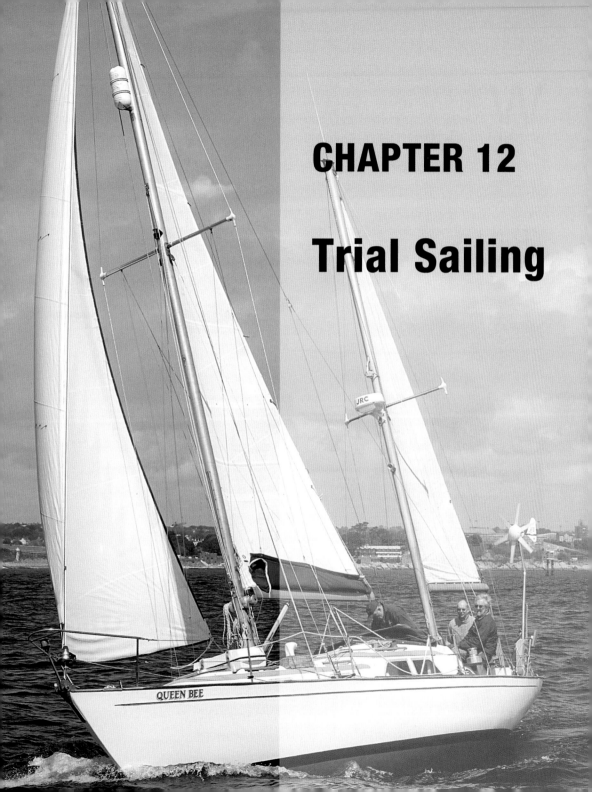

CHAPTER 12

Trial Sailing

When I trial sail yachts for magazine reviews I force myself to be objective about the boat, despite what my personal feelings might be. I recommend that you try to take a similar attitude when you first try out the boat you might buy. Try not to let glossy woodwork or fancy gadgets sway your judgement. Instead, approach the sail trial in a methodical and analytical manner, as it will be far more useful when you have to make the final decision.

Take a notebook, pencil and camera to record your feelings as you go along and take the time to jot things down as they occur, otherwise you'll get back home having forgotten half of what you discovered about the boat – and most likely all the bad bits! The mind of a boat buyer is very subjective if he's in love with the thought of owning that particular yacht, so if you don't jot down the not so good things about it right away, there's every chance you'll gloss over them when it comes to weighing up all the pros and cons.

Keep good notes when you're test sailing her or you'll soon forget things

Also, insist you take a full part in the running of the boat – you'll learn little if you let the current owner run around doing everything while you simply steer her.

Testing a Used Boat

Under power

Start by putting the engine into gear whilst still firmly attached to the pontoon. Take a look at the exhaust to see if there is any black smoke – often an indicator of injector problems in diesels. Also, make sure there is a good 'whoosh' of cooling water coming out – a weak trickle usually signifies a worn water pump, impeller or similar. This is equally important with a petrol outboard, which should have a telltale cooling water outlet behind the drive leg.

Now have a look over each quarter to see which direction the prop wash is going. Try a few revs astern and see if you can get a feel for where she wants her stern to go. The more you know about how she's likely to perform under engine before you set off, the better. Does she have a Saildrive? If she does it's likely there'll be a bit of a pause between putting her in forward gear and feeling some steerage through the rudder. This is because Saildrives tend to be a fair way forward of the rudder, so any prop wash, the effect of which you'd feel immediately on a shaft-drive boat, takes its time. In effect this means that giving her a short, sharp burst of power ahead won't always nudge her stern around as quickly as you might like.

Most handling problems experienced in a boat under power can be overcome, or even used to your advantage once you become used to them. I've owned long-keelers in the past, with offset propellers as well, so quite where you were likely to end up when going astern was anybody's guess! However, after a year of squeezing her in and out of awkward marina berths, locks, canals and harbours, we were soon able to predict her rather erratic behaviour to within a foot or so and ended up using her quirky manoeuvrability to our advantage.

Prop walk is the way a propeller acting on the water causes the stern of the boat to move in one direction or another – and is visible and detectable by putting her into reverse and giving the engine some good revs. Once you know that she pulls hard to port when revved hard in reverse, then you'll do your best to favour portside-to berths so that when you apply loads of revs astern to stop, she will neatly pull herself alongside.

Once away from the mooring into open sea, try spinning her in her own length and seeing how quickly and tightly she'll turn. Remember to do it clockwise and anticlockwise as, depending on the action of the propeller, she'll turn tighter in one direction than the other. A trick worth learning is that a yacht will turn quicker if you put her into neutral halfway through the turn, as this eliminates the adverse effects of prop walk and allows her to spin around her keel. Usually, the shorter the keel the quicker she'll turn and in the least amount of space.

Prop walk can make the stern of the boat move one way or another

Under sail

Once you're happy that the engine is working okay and that she handles reasonably well, it's time to hoist some sail. I always start with just the mainsail, to give me an idea of how she copes under main alone – what speed she'll attain, whether she will tack okay, how much **weather helm** she shows and how much of a tendency she has to round up. This way I learn how much work the headsail does and how much it can be relied upon to adjust the balance of the boat.

These days most new production yachts have fractional rigs with large mainsails and small jibs, so the mast should be positioned slightly further forward to compensate. In other words, she shouldn't necessarily be craving to round up under main alone and should still show a good turn of speed without the jib hoisted.

See how well she handles in the open channels before attempting to berth her

Take a good, close look at the sail to see if it is worn, damaged or stretched in any way – particularly around the main stress points such as the clew and reefing points. Most old sails will have stretched somewhat and this can be seen by the extra bagginess in the draft of the sail, despite tightening the outhaul and kicker. Ideally with both of these hard on in a good blow the sail should be quite flat. If there's an adjuster on the backstay crank it up a bit to see if this makes much difference – if not, then you might need to make room in the budget for a new mainsail.

Hoist, or unfurl the jib and make sure all the running rigging is in order. Once again, check the jib for wear or damage in all the usual weak points and go up forward to take a careful look at how the sail is pulling and whether it is creating a good slot between it and the main. Make sure the traveller is in the correct position for the cut of the jib – i.e. roughly bisecting the angle from the clew – then see if the leech of the sail flaps at all even after tightening the leech line. If it does, or the sail is inordinately baggy, then you might need a replacement jib as well!

Seeing how she sails under mainsail alone can be quite useful

Take a look at the standing rigging as well when you're sailing, particularly at the leeward shrouds. While you wouldn't expect them to be as taut as the windward ones, neither should they be flapping around loosely. Also, have a look up the mast to see if it is bending out of shape under the loads – if it's out of true vertically, in a sideways direction, then the rigging is not set up properly and will need to be re-tensioned.

Although fast becoming a dying art, heaving-to can tell you a lot about a boat. In the days of long keels, heaving-to was a fantastic way of riding out a storm in a modicum of comfort and safety. With modern fin and skeg hulls it doesn't work quite so well, but it's worth giving it a try to see if it calms things enough to make a cuppa, or go forward to fix something.

Heaving-to is achieved by simply tacking and leaving the jib backed – i.e. leave the jib sheets where they are initially. Once through the tack, the idea is to stop the boat in the water and leave the wind somewhere around 50–70 degrees off the bow. This is achieved by carefully balancing the amount of power trying to turn the boat into wind (from the mainsail) against the amount the backed jib is forcing the bows away from the wind. The helm should end up fully held over towards the wind – with a tiller that will be fully down to the leeward side, with a wheel wound round as far as you can go towards the wind. Lock the helm in this position so that, in effect, you will be stalling the boat out. With any luck she will make no more than a half-knot of progress through the water and at a comfortable enough angle to the wind and waves to allow you to move about safely on the boat. You might find you will need to reef one or other of the sails to obtain a good balance, especially if you have a large genoa out, and you might need to 'bag' the sails out a little by letting off the main outhaul or windward jib sheet.

Always go below when you're under sail and listen out for unusual noises and creaks. Some may be just a bit annoying, but others may foretell of problems to come. Open and close any doors, especially those near the mast, to see if there is any give in the mast step that is forcing the superstructure slightly out of shape. If this is the case, then you'll need to take a closer look at the mast-step area for any stress-crazing or cracks as it might be in need of reinforcement.

Good sails are essential for getting the very best performance out of her

If she's a cruising yacht, while you're below lie in a few bunks to get a feel for how she rides through the water and where the best sea berths are. Stand at the galley and imagine you're preparing a meal and try generally moving about under way to see if she has sufficient handholds in the right places. Also, lift a few sole boards and check again for any incoming water – some leaks may not show up until the yacht is subjected to the typical loads endured when under sail. Another essential to a cruising boat is the heads, so try it out. Does it flush properly?

Heaving-to will give you time to stop and think, or to carry out a quick repair

You're unlikely to be able to choose the weather for your test sail, but, within reason, don't be put off going out in a strong blow – you'll most likely learn more in testing conditions than you will in a gentle breeze.

Testing a New Boat

When test sailing a new boat the actual process will be very similar, except that all the equipment is brand new and should, therefore, be in tip-top condition. So, if the sails don't appear to trim correctly or any of the deck gear doesn't perform as it should, you need to ask the dealer's representative on board if this is how it normally is, or if there is a particular reason for the malfunction. If he can show you where you might be going wrong in setting it up for best performance, then all well and good – you'll learn something and will get to see the boat sailing at her best. If he can't, then note it down to ask the builder, designer or owner's association when you get back.

Production boats built down to a price will most likely be equipped to the minimum acceptable level in regard to such items as winches, blocks, clutches, travellers etc., so it's a question of weighing up the cost savings against the type of sailing you plan to do. If you make sure you try out all the elements of sailing her, especially handling the sail control lines, you'll be able to judge if any of the deck gear might need upgrading to a more powerful system. If you aim to sail offshore, or in open ocean conditions, you'll probably want to take the deck gear on a standard production yacht up a level or two, so make sure there is the option to do so as not all boatbuilders are flexible in this matter and retrofitting deck gear will be inordinately expensive.

In a used boat you might forgive the odd hatch drip or bulkhead creak, but with a new boat everything should be 100% correct and working as stated. But bear in mind that problems do arise whatever age the boat is, so don't be put off from buying a boat because of a misplaced 50p seal. All the same, if you detect even the smallest fault it's worth noting it down and making

Make sure all the sail controls and deck gear are up to scratch

enquiries – it might be a common problem with that particular model and if it is you'll be in a much better bargaining position before you've signed the deal than after.

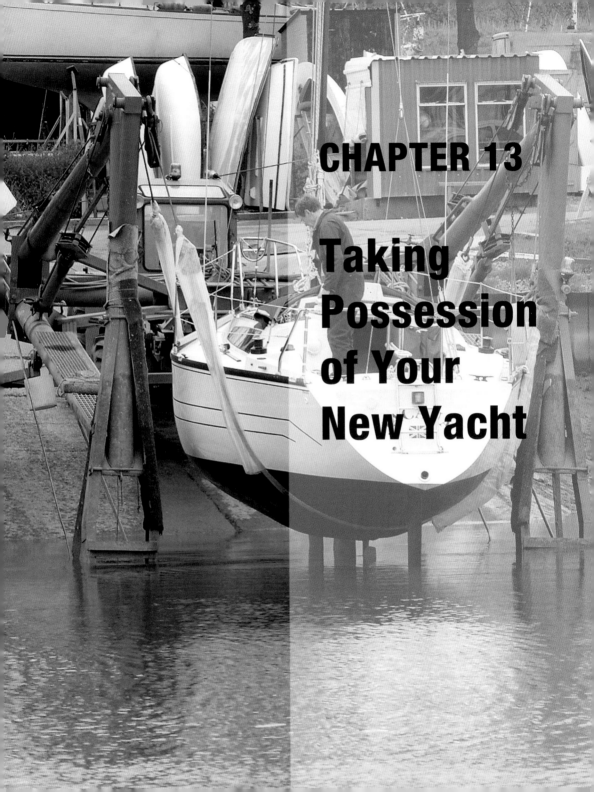

CHAPTER 13

Taking Possession of Your New Yacht

New Boats

Once you've decided on buying a new boat most builders/agents will take an immediate deposit, usually refundable if you change your mind after a trial sail, but double check first. If all goes well on a trial sail, then you must come to an agreement as to how you intend to pay, which will form the basis of a contract between you and the builder. This legally binding contract will contain details such as the deposit amount, date and amount of the stage payments required during the build (if required), and the final delivery date. It should also contain details of your legal rights in the event of the builder going bankrupt during its construction, together with the list of extras ordered and all the warranty particulars.

Make sure you take legal advice on the purchase contract before signing

Whether you're buying a new or used yacht try to ensure that any deposit or interim payments you make to the builder or broker are put into a client account that will protect your money should the agent go out of business. This is hard to ensure because in the past some failing dealers have been known to say one thing, but actually do another. Their customers, thinking their down payments were safe in a client account, found they lost everything when the company collapsed not long after. Take legal advice and, if you can, set up your own managed client account so that money is only transferred to the builder/dealer after close scrutinisation by a third party. While it might cost a little more this way, it could end up saving you thousands if the builder goes into liquidation halfway through producing your boat.

Also insist on retaining the final balance of payment, normally around 10%, until handover day, when you are completely satisfied that the boat is exactly what you ordered and is fully functional. If you are in doubt about anything in the contract, either contact a lawyer with marine experience to check it out, or if you are a member, speak to the legal department of the Royal Yachting Association (RYA). The RYA has an excellent legal department so it might well be worth joining before you even start looking for a boat.

Commissioning and handover of a new boat

Commissioning can be one of those mysterious 'grey areas' to new boat buyers and is often a rather hefty sum tacked onto the end of the price list. It should entail a scrupulous inspection of the vessel and rig and a thorough test of all her systems, equipment and instrumentation to ensure they are set up correctly and functioning properly. It usually takes a few days, and involves tuning the rig, bending on the sails, installing instruments and running the engine, pumps, lights, cooker etc.

Once the yard has completed the commissioning you *must* give her a full sea trial, and check that all the equipment

A happy moment when the agent hands over the ship's papers

works yourself. Only after this should you hand over the final payment and sign the acceptance papers, in return for the warranty certificate.

At the conclusion of the sale the agent should furnish you with a Builder's Certificate, VAT receipt, proof of conformity to the RCD, registration document (if requested), warranty documents and operating/service manuals for the boat and all the equipment therein.

Used Boats

Beware! This is the point when you're most likely to let your heart rule your head. In a private sale, without the guidance and experience of a broker or builder, it can be all too easy to make a mistake that you might well regret later!

Agree in advance who pays for any hoisting out or launching fees

Always make your initial offer subject to survey (assuming you want one), so if problems are discovered during close inspection that either the surveyor or you feel must be remedied, you are entitled to reduce the offer by the cost of any necessary corrective work once you have obtained a few reasonable quotes for the work required.

Be wary of accepting any survey that the owner might proffer (it is not unheard of for vendors to 'remove' offending pages on the copy) and disregard 'insurance surveys' as these are merely valuations rather than full condition surveys and will tell you very little. A copy of a relatively recent condition survey can indeed be useful, particularly if it shows up problems that were found and subsequently sorted out. You will need to see copies of invoices, though, showing the work was actually carried out by a reputable yard or craftsman. In my experience, if the vendor has a thick bundle of old invoices, mooring bills etc., which he is happy for you to browse through, then the seller is usually pretty genuine and likely to have maintained her to a fairly reasonable degree.

Finally, if she is out of the water when you make your offer, let the vendor know (and write it into the Bill of Sale) that you will be holding back some of the money (around 10% is common) until she is launched and the engine has been run and tested satisfactorily. Alternatively make it part of the deal that you want her launched for a sea trial before parting with any money, although the vendor might baulk at the cost of a two-way hoist should you change your mind once she is launched.

Delivery

The first choice for many owners is to sail their new boat home and obviously, if she is near to the port you intend keeping her in, this makes a lot of sense. It is essential, however, to give her a thorough sea trial before embarking on a long passage. Common problems, especially with a boat that has been laid up for a long time, are engine failures due to water/muck in the fuel tank, rig failure, seizure of seacocks, and flat batteries or other electrical problems caused by corrosion and lack of use. According to the Royal National Lifeboat Institute (RNLI) a large number of distress situations arise during delivery trips, so do check her out thoroughly before leaving.

Make a list of everything that might need checking or servicing and work your way through the list methodically, leaving nothing to chance. If the boat has been unused for a while, check every seacock, hose clip and hose. Drain the fuel tank, replace the filters and refill with fresh fuel. Change the oil and filter. Check the batteries and, if in doubt about their age or condition, replace them. Check the navigation lights, VHF and instruments, and pack a spare handheld VHF and GPS just in case.

After checking the boat over thoroughly, ensure you have all the personal safety gear you might need – lifejackets, harnesses, liferaft (hire one if necessary), handheld GPS & VHF and a comprehensive first-aid kit.

It is well worth carrying out a thorough rig check before departing

If you're concerned about any safety aspect of your new boat contact the RNLI and ask them to send one of their staff to give her a 'SEA Check'. This is a completely free, friendly and confidential service that looks at safety aspects involved with your boat. While it is not actually a test or an official inspection, SEA Check is a personal safety advice service that takes place on board your own craft. Wherever you live in the UK or the Republic of Ireland this important free service is available to anyone who goes to sea in almost any type of leisure craft, so do call them on 0800 328 0600 or apply online at www.rnli.org.uk.

Professional delivery

If your newly purchased yacht is a long way from her new mooring, then you might prefer to have her delivered professionally – especially if it involves a long passage, Channel-crossing or

The RNLI are happy to carry out free SEA Checks to ensure your yacht is safe

similar. If you know of a skipper with a particularly good reputation, who you would entirely trust with your pride and joy, then go ahead and ask him, but check with the insurance company first – they might want proof of his credentials as a skipper. Otherwise delivery agents and crews can be found advertising in most sailing magazines.

If you have bought your new boat from a broker tell him what you want to do – he will most likely know of a reliable delivery skipper or agency. If you don't know someone personally, it is often better to choose a larger company to do the job than a single skipper, as the former will be well covered by insurance and familiar with the vagaries of organising flights etc. if necessary. Furthermore, many of the larger agencies will be able to combine different modes of transport, say, by taking her half the journey by sea and the other half by road.

Many delivery skippers will be happy to have you along for the experience, but don't expect a discount for your contribution. Some will even give you a little informal instruction on the way, although this is by no means part of their job description and cannot be assumed unless you have specifically requested it. No matter how qualified you are – the delivery skipper will always be *the* skipper throughout the trip and will be making all the decisions. Your job will be as crew only.

When initially enquiring, ask for a final, fixed fee that covers all costs including flights (if buying abroad), crew, food, fuel, charts etc. Don't be afraid to check out the skipper's references or speak to previous clients and make sure you inform your insurance company.

A delivery agent will need as much information about your boat as possible, including a detailed equipment inventory, most recent survey, registration document and insurance details. The same checks as you would do yourself will be carried out by the delivery skipper, regardless of whether you have already done them, and this is all included in the price. The skipper will be much happier if you, or the broker, are with him during this process as problems can usually be sorted out straight away without lengthy telephone calls. It is vital to tell the absolute truth about the boat's condition so that the skipper can arrive fully pre-

Getting her home safely might mean hiring a delivery skipper

pared for any work that might be needed before departing. Hiding something from him can not only put the delivery crew and your boat in peril, but also create delays – for which you will be charged.

By road

It can often make more sense to have your new boat delivered by a low-loader truck from another part of the UK or continental Europe. It is usually safer, saves wear and tear on your new boat and means you can carry out sea trials at your leisure in your own sailing area.

If it's a long way by sea why not consider moving her over land?

There are a good many agents that specialise in boat delivery by road who will be happy to quote you a price including craneage, mast unstepping, delivery and storage ashore or relaunching. Most are able to co-ordinate the whole procedure including hauling her out/back in and will give you a price for the entire operation.

As with any crossing of a national border, deliveries from another country will need the requisite paperwork to be given to the agent, including registration documents, VAT receipt and insurance certificate. Given the necessary documents, though, the transport agent should take care of all Customs and Excise formalities without you being present.

If you can allow the agent some flexibility with the delivery date, you might be able to get your delivery done as a return load, which will decrease the price considerably.

CHAPTER 14

Can I Share a Boat?

Compared with owning a boat outright, sharing is considerably more economical and can have other benefits such as sharing out the tedious chores like antifouling, engine servicing etc. If you are friends with the other partners, then it also gives you someone else to go sailing with on occasions when you don't have a crew to hand.

Alternatively, organised schemes whereby you simply 'turn up and go' take all the hassle out of owning a boat and are ideal for those with limited free time or who just aren't interested in the nitty-gritty of boat maintenance. It's a bit like getting your car valeted, rather than washing it in the drive on a Sunday and chatting to your neighbours at the same time. Though more expensive than a simple syndicate, compared with outright ownership these schemes can offer good value for money. The main downsides are that you miss the marina/club banter as you carry out the winter jobs and you are tied to where you can go and when, so the end of each allotted sailing period will be like getting a charter boat back to base on time. You won't just be able to leave her somewhere and pick her up again in a few weeks, instead you'll have to battle back against the wind and tide unless the next clients are happy to pick her up where you leave her.

Private Syndicates

A popular way of purchasing a boat is by sharing the cost with one or more interested friends. Many owners use this method to enable them to sail a much more luxurious yacht than they could afford alone. Not only does it reduce the initial outlay, but also, and often more importantly, it reduces the individual shareholder's annual mooring and maintenance fees to a fraction of what they would be under sole ownership. This means you might be able to afford to berth her in a marina, rather than on a swinging mooring, and it can make the difference between taking on a professional to maintain her, or doing it all yourself.

Difficulties can arise when one partner needs to obtain credit for his share, though, as a marine mortgage is only usually available for a shared boat when all the partners sign up to the deal – offering the boat as security. Obviously there are no such problems with an unsecured loan.

There are several ways to set up a private boat-share syndicate. If you have a friend or two interested in joining you, you can set up a partnership using a simple contract obtainable from the RYA. Any agreement should cover all details and eventualities including:

• Names, contact details and signatures of each partner
• Size of share owned by each partner
• Details of the boat and its mooring
• The name of the syndicate manager and his responsibilities
• Methods of disposal of shares in the event of sale or death
• Procedure in the event of a default by one partner

- Insurance details
- Registered skippers and their competence level
- The division of time allotted to each member
- The procedure for changing time allocation
- The condition to leave the boat in
- Running costs and the provision of a maintenance fund
- Voting rights of members to alter any terms in the agreement
- Dates of syndicate meetings

The syndicate should appoint a secretary annually to administer the general workings and maintenance of the boat, including paying mooring dues etc., and agree on who should carry out specific tasks on the boats such as cleaning, winterising, antifouling etc. at the outset. It is advisable to hold a members' meeting at least once a year, to decide the upcoming season's sailing allocations. Flexibility is the key to any partnership's success, so, if something needs fixing while you're using the boat, ideally you will fix it and discuss cost/responsibilities later.

Many casual syndicates have been successful, but there are those that fail

On withdrawing from the syndicate, a member would usually offer his share to the remaining partners prior to offering it for sale elsewhere. It can sometimes create ill feelings, but the remaining syndicate members must have the right to reject any potential partner they deem unfit for any reason. Sometimes an existing shareholder will offer to buy up any spare allotment, giving himself a larger proportion of the boat and the time allocation for sailing. However, in return for twice the sailing time, they would also have to pay twice the upkeep costs.

Plenty of owners have operated very successful private syndicates for years and swear by them. Others have not been so lucky. The fewer there are in the scheme the less likely you are to fall out. Either way, you should always consider the worst-case scenario when you are first setting up a syndicate and creating the rules. A regular cause of failure of such schemes is when one of the partners has damaged the boat and not informed the others, or one leaves the boat in an unseaworthy condition so that another spends most of his weekend fixing rather than sailing the boat. Another problem is when one of the partners hits bad times and can no longer afford the latest toys and goodies that the others in the syndicate want to splash out on. You might be happy to get along with paper charts and warm beer, whilst the others favour the latest colour chart plotter and a new fridge.

As an alternative to setting up your own scheme with friends, colleagues or acquaintances you could use a professional boat-share agent to find you a share in an existing syndicate. They will sort out all the necessary paperwork for you and introduce you to the others in the partnership. This method also offers the benefit of a greater choice of boats and homeport locations.

Some very keen sailors have even been known to join two or more syndicates in different locations to stretch their sailing season and vary their cruising grounds.

To sum up the pros and cons of shared ownership then:

For

- Newer or bigger boat for your money
- Better equipped boat
- Reduced costs for mooring and maintenance
- Maintenance tasks shared
- The boat remains in constant use

Against

- Possibility of falling out with partners
- Tied to specific dates and periods of sailing
- Possible difficulty in selling your share
- Commitment to spend on new equipment

Fractional Ownership Schemes

Over the past decade there have been a number of developments along the lines of boat charter, but with more of a 'club' or 'syndicate' arrangement. These commercial schemes are often referred to as 'boat-share', 'subscription' or 'fractional' sailing and usually allow you a fixed number of days or weeks of sailing each year, either on the same boat or one of a pool of similar boats. Unlike a charter/partnership scheme, you don't own any part of the boat you sail, but merely rent time on board rather like a standard charter.

If you are retired, say, and have plenty of time to sail, then owning a boat outright might make good sense. But if you work and can only take three or four weeks off a year and a few weekends, it might well be more economical to join one of these schemes.

One of the first companies to introduce such a scheme, SailTime, started in the USA around 15 years ago, and has now been operating in the UK for 10 years. It already has a dozen or more bases in the UK and is expanding rapidly into Europe. With SailTime you can either be an owner-member by purchasing a boat to the company's approved specification, or you can simply take out a monthly membership entitling you to sail.

Sometimes joining an organised fractional ownership scheme is better

Owner-members are paid a guaranteed sum each month, plus they are entitled to as much sailing time as the other members of the scheme. Sail-only members take out a contract for a minimum of 12 months, entitling them to an eighth share of the sailing time on a particular boat, so when booking you only ever have a maximum of seven others to negotiate with.

SailTime boats are all reasonably new and far more comprehensively equipped with navigation electronics etc. than the average charter yacht. They are available to sail all year round, except for a two-week period in January or February for annual maintenance (the cost of which is all included in the one monthly payment).

Bookings for sailing allocations are made remotely via the Internet through a sophisticated booking system and your choice of times is obviously controlled, so as not to deprive other members of their fair share. Sailing periods are divided into what are called 'Sail Time' slots, which are between the hours of 1000–1830 and 1830–1000. Each member is entitled to seven of these Sail Times per month and they can be booked sequentially to give up to a week on board in one go.

An on-line scheduler allows you to book a half-day, full day, weekend or a full week up to a year in advance. One whole week will use up two month's allowance, but this can be supplemented by an unlimited amount of 'Anytime' sailing, booked a maximum of 24 hours in advance, where un-booked slots are still free. These might not necessarily be free at the end or beginning of your booked week though, so the longest amount of guaranteed continuous sailing would be one week. All skippers in sole charge of a SailTime boat must hold a Day Skipper certificate at least, or else undergo training with SailTime's own instructors.

A few other fractional sailing organisations exist that are run in a similar fashion, although maybe not quite so slickly. That said, SailTime is the most expensive of these organisations, with a one-off joining fee as well as a monthly membership fee.

For

- No mortgage repayments
- No mooring fees
- No maintenance charges
- No insurance costs

Against

- Limited time to sail each month
- No capital share of the boat
- Joining fee

Charter/Purchase Schemes

Another less costly way of buying a new boat is by using one of the various partnership schemes run by many of the larger charter companies. However, it really only makes economic sense if you have the time to make the most of the free perks, such as using the boat when it is not being chartered and hotel and/or flight discounts to other destinations.

Offers vary depending on how you intend to use the boat and how much you wish to invest. You can either buy the boat outright and then lease it to a charter agent to get an income when you are not using it yourself, or you can choose a partnership scheme whereby you pay a percentage of the cost price of the boat up front – usually 50–65%, which gives you the right to use your boat, or a similar one, for several weeks a year, whilst the company charters it out for the rest of the season and gives you a percentage of the income.

Charter-to-own schemes can save money and give you plenty of sailing

During the contract period (usually three to five years) the company berths, cleans and maintains your boat at its own expense. At the end of the contract the boat is yours to keep, or you can part exchange it for a new boat and start all over again.

Larger companies will often organise the purchase, commissioning and delivery on your behalf and some can even provide finance on the remaining balance.

For

- Mooring fees, insurance and maintenance paid
- Guaranteed income
- Personal holiday benefits in various locations

Against

- Boat suffers considerable wear and tear
- Often requires new sails and/or engine on return
- Boats have to be the charter company's preferred model and specification

CHAPTER 15

Documentation

S/Y Olivia

Vessel

Details

Type	Nicholson 35
Launch date	1971
Designer/Builder	Camper & Nicholson
Hull No.	35/003
Hull length	10.76m
Beam	3.20m
Draught	1.70m
Displacement	8,013kg
Engine	Bukh 36hp (26.5kW)
Berths	6
Registration No.	SSR84036
Country of registration	UK
Home port	Portsmouth
VHF Radio License No.	32026A

Registration certificate

CERTIFICATE OF BRITISH REGISTRY
REGISTER OF BRITISH SHIPS PART III

DETAILS OF SHIP

DESCRIPTION: SAILING YACHT
OVERALL LENGTH: 10.76 Metres
NAME OF SHIP: OLIVIA

S/Y Olivia

Other equipment

Electronic navigation

Item	Model
GPS main	Furuno GP1850
GPS hand held	Garmin 45XL
Chart Plotter 1	Furuno GP1850
Chart Plotter/Radar	Raymarine E120
Radar	Furuno 1712
Navtex	Furuno NX-300
Autopilot	Autohelm 4000+
Sailing instruments	Raymarine ST50

Electrical

Item	Model
Service batteries	4 x 90Ah Trojan 27RVX
Engine battery	1 x 100Ah starter
Mains battery charger	Sterling 30A 3-stage
Wind generator	Ampair Pacific 100

Other

Item	Model
Inflatable dinghy	Avon Redstart
Outboard engine	Yamaha 2B 2-stroke
Anchors	1 x 15kg Bruce
	1 x 15kg Fisherman
	1 x 8kg Danforth
Chain	Bower - 50m x 8mm
	Kedge - 10m x 8mm
Bower chain markings	10m - 1 Gn
	20m - 2 Gn
	30m - 1 Rd
	40m - 2 Rd
	Also painted red every 15ft

S/Y Olivia

Safety

Equipment

Liferaft		
	Type	Avon 4-man
	Serial No.	40930
	Service expiry date	
Lifejackets		6 x auto inflation
Harnesses		6 x integral with lifejackets
Life rings		2 with ship's name
Flares		
	Red para	4
	Red hand	4
	White hand	2
	Orange smoke	2
EPIRB		
	Type	McMurdo E3 406MHz (Cat 2 Manual)
	Registered	Yes (232G Britain)
	Serial No.	9D0E1A870800001
	Battery expiry date	
PLB		No
SART		No
AIS		No

Liferaft certificate

SEE OVER

Yachts are unusual possessions in that they are highly portable and can also be very expensive. They are also potentially dangerous and contain specialist equipment such as marine radios, satellite phones etc. All this leads to the need for regulation and inevitable documentation.

The requirements change over time and from country to country, with the official line sometimes differing widely from what is actually enforced. Your national sailing body should be able to inform you of the latest rules and regulations. The information that follows focuses mainly on the UK.

Documentary Requirements

Registration

You are not legally bound to register a UK-owned vessel under 300 **GRT** (Gross Registered Tonnage) that is only being used in UK waters. However, sailing in foreign waters in a boat over seven metres long requires proof of ownership and registration. Also, the boat must fly the ensign of its country of registration as well as a courtesy flag of the country whose waters it is in at the starboard spreader.

There are two forms of UK registration for pleasure craft – Part 1 on the Register of Shipping, or Part 3 on the Small Ships Register (SSR) for boats under 24 m (part 2 is for fishing vessels only). The former is indisputable proof of title to your vessel and allows you to choose and retain use of a unique boat name. The boat, new or old, has to be officially 'measured' for its tonnage, for which there is a charge of around £250, plus a further sum (currently £124) to renew it every five years, or for any detail changes such as new engine etc. You must be a British citizen, or a citizen of a British dependency to hold Part 1 registration.

Registering a used boat on Part 1, however, requires you to prove ownership by bill of sale and that the vessel is free from debt. Once completed, it allows you to register a mortgage on the vessel, which is declared as a 'lien' on the registration document.

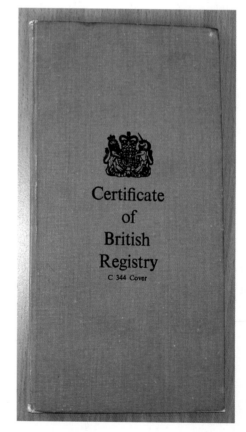

The Blue Book signifies full Part 1 Ship Registration

If you are buying a new boat it is advisable to go for Part 1 registration as it is quite simple at this stage and often helps when you come to sell it, rather like the service history on a car. It will be more expensive than a simple SSR registration, though.

Part 3 (SSR) registration lasts for five years at a time and does not constitute proof of ownership. However, many boat owners choose this form of registration as it is considerably cheaper than Part 1 and is generally accepted all over the world as proof of the vessel's home port.

Both types of registration entitle the holder to wear the red ensign only. Blue ensigns are for retired naval officers and members of a 'Royal' yacht club, whilst white ensigns are worn on serving naval vessels. Some clubs are entitled to fly defaced ensigns with royal permission.

To register contact the Registry Of Shipping & Seamen, PO Box 420, Cardiff, CF24 5XR (Tel: 02920 448800), or go online at www.mcga.gov. uk for immediate SSR registration.

All British yacht owners are entitled to wear the red ensign

Bill of sale

A bill of sale (BOS) is a document that is used to transfer title of the vessel from one person to the next. An official blank form is available from the Maritime and Coastguard Agency (MCA) that must be used for Part 1-registered transfers, but can also be used for SSR-registered used boats, either bought from a broker or sold privately.

The BOS contains details of the old and new owners and a signed declaration that the vessel is free of any liens (mortgages etc.) at the point of sale.

VAT

When buying or using a vessel in EU waters it is now essential to have some proof of the Value Added Tax (VAT) status of the vessel, especially when travelling to another EU country. Boats launched after 1 January 1986 must have the original invoice showing the VAT paid. Those built

A blank bill of sale can be downloaded from the MCA's website www.mcga.gov.uk

prior to that date are exempt from paying, but must carry proof that they were in EU waters before that time – i.e. a bill for mooring, servicing or similar, prior to that date.

Many foreign registered and ex-charter boats are put on the market without the VAT being paid. Buying one will render you liable for VAT on its estimated value (inspected and valued by an HM Customs' appointed surveyor) at the VAT rate levied in your chosen port of registration. If you pay it in a country where the VAT rate is lower, you will be liable for the difference if ultimately importing the yacht into the UK. Boats registered outside the EU and owned by a non-EU resident can visit any other country in the EU for six months only (unless laid up out of the water and bonded by Customs) without paying VAT, but the yacht must then leave the EU completely after this period.

It is possible to claim back VAT on a new boat if you export it to a non-EU country within three months of purchase, but the VAT must be paid if it is subsequently sold on to a permanent resident in the EU.

EU Recreational Craft Directive (RCD)

All boats over 2.5 m (8 ft 3 in) LOA built in or imported into the EU after 16 June 1998 must comply with the RCD and display proof of same on a certificate and a hull plate. This often makes it uneconomic to import a boat from outside the EU, regardless of its age, as you will be liable for bringing it up to the relevant standards of the RCD in force at that time.

There are a few odd exclusions, namely classic boats of a pre-1950's design, racing boats, self-build boats and vessels that originated in the EU prior to 16 June 1998, but were exported and then returned.

Excellent advice on the VAT status of boats is available from the RYA's legal department on 0844 5569519 (you must be a valid member), or CE Proof on 01603 717181, www.ceproof.com.

Radio Licences

Although not compulsory for leisure craft below 300 GRT, having a VHF

VAT Payment Certificate

Relating to
Boat type:

Hull No:

Hull Identification No: CE HIN:

Name of Yacht:

Present owner:

Address:

The above mentioned yacht was built in the year: **2008**

by builders: **NORTHSHORE YACHTS LIMITED**

to the order of :   **AS ABOVE**

Our records show that VAT at the rate ruling at the time was paid on the full invoice value.

Our files are available for inspection by HM Customs and Excise if required.

Signed: Date:

For Northshore Yachts Limited

Registered Office:
Northshore Shipyard, Itchenor, Chichester, West Sussex PO20 7AY

It is essential to have a VAT receipt if the yacht was launched after 1 January 1986

radio on board and knowing how to use it is an essential component of the vessel's safety equipment. Very few cruising yachtsmen would consider going to sea without at the very least a hand-held VHF radio and most consider a full power, fixed VHF radio as par for the course.

All new marine VHF radios are now equipped with Digital Selective Calling (**DSC**) and single-button distress calling to comply with the Global Maritime Distress and Safety Scheme (GMDSS). On issuing of your Ship's Licence you will be given a Maritime Mobile Service Identity (MMSI) number to enter into your VHF that is similar to a mobile telephone number and identifies you and your vessel to the emergency services. It also allows you to receive direct calls from other vessels without broadcasting on VHF Channel 16.

Installation of a maritime VHF radio into your yacht will require you (a) to obtain a Ship's Radio Licence and (b) to have at least one member of the crew who holds a Short Range Certificate (SRC) operator's licence.

Ship's Radio Licence

Licensing ensures that radio equipment used on board ships does not cause undue interference to other communications equipment and is operated by competent persons.

Licensing details, including the **call sign** of the vessel, name of licensee and other information are sent to the International Telecommunications Union (ITU) and MCA and could be used to assist in co-ordinating rescue operations if required.

If you own a vessel and you have radio equipment that is fixed to the vessel you will need to apply for a Ship Radio Licence. If you intend to use a handheld radio as well as fixed radio equipment on board your vessel, this can be registered under the same licence application as the fixed equipment, but it will be licensed for use on this vessel alone.

The terms and conditions form part of the licence and should be printed out and kept with the licence at all times.

If you wish to use a handheld radio only then you will be required to apply for a Ship Portable Radio Licence. Instead of a call sign (applicable to vessels) you will be provided with what is known as a T-number, which will enable you to use your portable radio on board any vessel.

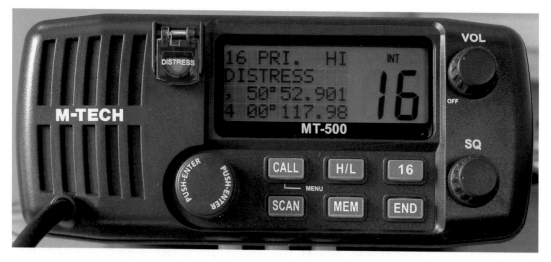

In my opinion every sea-going yacht should carry a DSC VHF radio

Initial applications and renewal/changes can be made online at www.ofcom.org.uk or by completing a paper form and sending it to the licensing authority Ofcom, which will incur a small charge.

If you are buying a used boat and the vessel's licence lapsed more than five years ago, you will not be able to use the same call sign and MMSI because they are recycled every five years.

When you apply for your radio licence (using either the online licensing system or the paper forms), Ofcom will allocate you a new call sign and MMSI number. It is your responsibility to programme the new details into your radio apparatus.

Radio operator's licence

All operators of marine VHF radios must either hold an SRC, or be supervised by someone who does. To obtain an SRC you will need to do a one-day course at an authorised licensing class. The RYA has a full list of such schools and can recommend one near to your home or boat.

---

www.ofcom.org.uk   Of347   **Form for Application/Amendment/*Validation/Surrender of a **United Kingdom Ship Radio Licence or a **United Kingdom Ship Portable Radio Licence**

Please note that Ship Radio Licence applicants and Ship Portable Radio Licence applicants are encouraged to register and apply for and print their ship radio licences via the Ofcom website www.ofcom.org.uk. This entitles you to a free licence.

A separate form should be used for each type of licence. Only one licence can be applied for on each form.

Please refer to the 'Ship Radio Guidance Notes for Licensing' Of168a when filling in this form.

Further information is available in 'Ship Radio Information' Of19a.

All fields are mandatory unless stated otherwise. Exceptions are Title, Tel No, Fax No, and E-mail. However, failure to supply this information may hinder Search and Rescue operations and may also delay the processing of forms.

Please use BLOCK CAPITALS and black ink throughout this form and use a separate sheet of paper for supplying additional information, clarification, or if space is not sufficient.

Tick ☐ if you have attached a separate sheet(s).

General information

The use of all maritime radio transmission equipment on board a vessel must be covered either by a Wireless Telegraphy Act Ship Radio Licence or Ship Portable Radio Licence.

Before completing this form, please read the guidance notes.

Please ensure that this application form is completed fully. Your application may be returned to you if it is not.

Important

1 Maritime radio transmitting equipment may be operated only by or under the direct and personal supervision of the holder of the appropriate Maritime Radio Operator's Certificate of Competence and Authority to Operate. Details of these are available from the Maritime and Coastguard Agency (MCA). See 'Ship Radio Guidance Notes for Licensing'.

2 The radio equipment covered by this licence must meet certain performance requirements as detailed in Section 4 of the guidance notes.

The Data Protection Act

The information you provide in this form and any information submitted will be used by Ofcom for the purpose of issuing, amending, validating and surrendering a Ship Radio Licence or a Ship Portable Radio Licence. See Ofcom's website for our data protection statement www.ofcom.org.uk

The information held by Ofcom will be updated as appropriate by relevant numbering records provided by Inmarsat, Accounting Authority Identification Code records provided by the (Maritime Radio) Accounting Authorities and EPIRB/PLB records provided by the MCA.

This information may be provided by Ofcom to the International Telecommunication Union (ITU) and the Maritime and Coastguard Agency (MCA) to assist in Search and Rescue operations and to help ensure that the information held by the ITU and MCA is accurate and up to date. This information may also be provided by Ofcom to law enforcement agencies (such as the police) for crime and taxation purposes as well as the Maritime administrations MCA Channel Islands and the Isle of Man. We will keep your information confidential and will not disclose it to third parties, except where we are required to do so by law or where we have obtained your consent in advance.

Please note that in general, licensing details of Ship Portable Radio licences are not sent to the ITU.

*   Licensees are required to validate their details at least once every ten years unless they have notified Ofcom that their details have changed during that period.

** Including the Channel Islands and the Isle of Man.

Ref: Of347/Of.105_10/06

OFFICE OF COMMUNICATIONS

Page 1 of 12

A Ship Radio Licence can be obtained online at www.ofcom.org.uk

---

If you hold the older style operator's certificate you will need to upgrade to an SRC if you intend operating a DSC radio. This short course will give you all the latest information and procedures for operating a DSC radio with the dedicated VHF Ch70 distress alert button and associated safety messages.

EPIRBs and other radio equipment

Your Ship's Radio Licence will also cover any other radio transmitting equipment you might carry on board, such as RADAR, EPIRBs, PLBs, AIS etc., so long as they are included in the list when you apply for your licence, or the licence is amended on the date of purchase of such equipment.

Once again, this is a quick and easy operation online.

Insurance

Whilst it isn't compulsory to have a leisure vessel below 24m in length insured in the UK at the time of writing, it is highly recommended that you at least obtain third party insurance in case you damage someone else's boat, or injure a member of the public. Most marinas now insist on seeing evidence of third-party insurance before they will offer you an annual berth.

There are marine insurance specialists who will tailor a policy to suit your requirements. Certain things will increase the premiums – such as where you keep your yacht, your geographical sailing limits and at what times of the year you sail. The cheapest policies usually cover you to sail in UK waters from April to October, whilst being kept safely tied up in a marina with 24hr security when not in use, and in a cradle, mast down and on the hard standing during winter.

You will need a Short Range Certificate (SRC) to operate a DSC VHF radio

Longer passages can be covered by ringing the insurers and letting them know in advance, although for ocean passages they are very likely to demand that at least one qualified Yachtmaster is on board. Premiums increase if you leave the boat unattended on a swinging mooring (buoy), or you wish to sail throughout the winter.

Before insuring your boat with a company, check it is a member of the Association of British Insurers (020 7600 3333; www.abi.org.uk).

To reduce the cost and give you greater peace of mind, it is worth fitting a security alarm, gas detector with automatic shut-off valve and an automatic engine fire extinguisher. A sailing qualification also helps keep the insurance cost down, as does accepting a higher excess.

CHAPTER 16

Buying
Abroad

B uying an apparent bargain boat abroad can be very tempting, particularly when the exchange rate is favourable. But be warned – although it is perfectly viable to purchase a boat at a reduced price abroad, especially blue water yachts where the owners have had enough of cruising and want out, there can be a few difficulties to overcome that might make it not seem quite so worthwhile.

Problems Buying Abroad

Unpaid bills

Common problems experienced when buying privately abroad (apart from any language barrier) include unpaid moorings, service or VAT bills, incorrect registration details and a lack of any proper ship's papers. Any one of these can lead to the boat being impounded by foreign authorities as soon as you try to move her. It is also not unknown for someone who doesn't even own the boat to try to sell her, usually to try to recover an unpaid debt from the real owner.

If buying abroad make sure the ship's papers are in order

If you use the services of a local broker to sort out the sale, it will cost you a little but he will be familiar with both the paperwork and the language and be able to find out much more easily than you if there are any outstanding claims.

The best deals are frequently on British-registered boats abroad, as they are often fully equipped with expensive cruising gear. The paperwork is usually in order for them to have been allowed into the country in the first place, although you must check the VAT status because some owners buy a boat and immediately export it out of Europe to legally escape paying VAT. If this is the case you will be liable to pay VAT at the boat's current estimated (by UK Customs or equivalent) value on its return to the UK, or if you keep her anywhere in Europe. Some believe that you can simply take the boat to the EU country with the lowest VAT rate and pay it there, but the UK authorities won't necessarily accept this and might well revalue the yacht and hand you an additional bill – especially if they think you were deliberately trying to avoid paying UK taxes. VAT payments on used boats is a very grey area unfortunately, so you can't be absolutely sure how your particular situation will be treated and, in my experience, no-one in authority will be inclined to put anything in writing to you!

Has the yacht overstayed her welcome?

Another thing that is important to verify is that the boat has not overstayed her temporary import permit into the country where it lies. Some countries will allow you to outstay this fixed time period if the yacht is on the hard and bonded by Customs – i.e. not able to be used – but the rules tend to change from country to country, and worse still, some countries ignore the rules if it suits them, but tend to invoke them at will if you get the authorities' backs up for any reason.

If the yacht is in the water, one possible way of protecting yourself is to inspect and test sail the boat where she is, then, if you definitely want to buy her, offer to pay the delivery charges back to the UK and complete the deal once she is back home. That way the owner will be forced to ensure the paperwork is up to date. This won't always be possible, though, as some of these boats will have been put on the market due to financial hardship, and to pay delivery, fuel and insurance costs to cover this would not be viable for the owner. If the latter is the case, then the price should reflect this – assuming, of course, that you actually want to bring her to the UK and not continue sailing her from where she is currently lying.

EU Recreational Craft Directive (RCD)

Probably the most serious deterrent to buying a foreign-built yacht from outside the EU and importing her to the UK is the problem of making her comply with the EU Recreational Craft Directive (RCD). As mentioned in the previous chapter on documentation, but worth reiterating here – all boats over 2.5 m (8 ft 3 in) in length built in or imported into the EU after 16 June 1998 must comply with the RCD and display proof of such on a clearly visible hull plate.

Bring her home or leave her there for some foreign cruising?

If the boat was built outside the EU, not to RCD standards, and thus not duly CE plated, you will have to put her through what can be an expensive and sometimes very difficult compliance process when you get her back – regardless of her age.

If, however, the boat was built and used in the EU before 16 June 1998, then exported, she will be exempt – providing you have written evidence (mooring bills, service invoices etc.) to the fact that she was used in the EU prior to that date.

Excellent advice on the VAT and RCD status of boats is available to members at the RYA's legal department on 0845 345 0373 and many of the relevant FAQs are free to view on its website at www.rya.org.uk.

Always take a test sail and go over the yacht thoroughly before setting off for home

Inspection and delivery

Don't forget to budget for an inspection trip and the passage/transport home, or you might find the final expenditure is no less, and possibly even more, than if you had bought a boat in the UK. Depending on the value of the yacht you're interested in, some owners will offer to reimburse your travelling expenses to view the boat if you go on to complete the purchase, but it's always worth trying to see more than one boat if you're going to bother to go abroad.

Once you're happy with the boat and you've had all the paperwork checked out, ring around and find out how much it will cost you to get her home before you agree to buy her. It's no good picking up a bargain in Gibraltar if you've got to pay a delivery crew £6000 to bring her back to the UK – well not unless you've already factored that into your offer or the owner has already reduced the asking price accordingly.

Also, bear in mind all the effort and wear on the boat bringing her back home on her own keel. Might it be better to put her on a lorry instead? A bargain used boat that's been left uncared for over a few years might take a fair bit of work to get up to a seaworthy condition and you certainly won't want to discover a critical fault when you're halfway across the Bay of Biscay in a Force 8!

When editing a sailing magazine I was inundated with disastrous sailing stories (for some reason folk forget to write about all the good times) and many of the worst ones were delivery trips of over 100 miles or so, where the proud new owner hadn't realised just how poor the condition of his newly purchased boat was before setting off in her for home. Seized engines and bilge pumps were all too common, as were ripped sails, chafed running rigging, rusty or bug-infested fuel tanks and filters, contaminated water tanks and more.

I would always plan to spend a week or so thoroughly going over your new boat and to carry out at least a couple of short test sails before setting off on the long passage home.

Keeping Your Yacht Abroad

Some prospective long-term cruisers will be buying their yacht abroad with the intention of keeping her there for a few years of sailing, before returning home. Well, everything I've mentioned above will still be as relevant whether you plan to carry on cruising in that area, or sail home. But, in addition to confirming you have the correct paperwork for the boat, you must also find out what permits you may need for your chosen homeport. Although the majority of Mediterranean countries are now part of the EU, they all have their own rules about visiting yachts and crews. Some insist you buy a cruising permit each year, others don't really care so long as you're not flouting any local laws. I know of a friend who left his boat in a tiny, far-flung fishing harbour in one of the Greek islands for 12 years without anyone ever questioning his right to be there. His was a UK SSR-registered boat that he bought off another UK resident who had swapped life on board for a house on the island and a Greek wife. But things have changed now in the Med, so don't expect to get away with flouting the local bylaws without being fined – or worse, having your boat impounded by the authorities.

By rights, you won't be allowed to register a boat in another EU country unless you become a resident and nowadays, with many EU countries desperate to raise taxes in any way possible

The authorities could stop you leaving if there are outstanding dues on the yacht

(especially from foreigners, and even more so from apparently 'rich yachties'), you will need to keep your situation in order to legally keep and use your boat there.

The rules change every day and one of the best ways to keep in touch with any changes in local regulations is to go online and search for any 'liveaboard' websites or blogs from like-minded sailors in the areas in which you are interested.

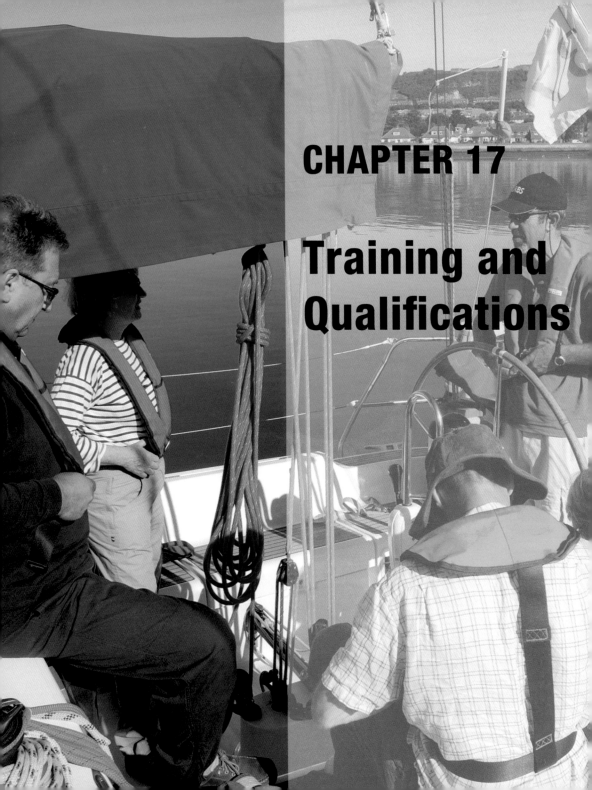

CHAPTER 17

Training and Qualifications

There are so many ways to get into sailing that it's hard to recommend an ideal route. The most common way to gain sailing experience, especially when I was learning some 40-odd years ago, is to work your way up through dinghies and day-sailers, learning from your mistakes.

However, if you prefer a more formal system, there's probably little to beat the Royal Yachting Association's (RYA) structured courses to take you from one level of competence to the next, slowly building up your knowledge base and confidence along the way. The RYA is the UK's national association for recreational boating and is currently the only training establishment recognised by the Maritime and Coastguard Agency – the UK government organisation formed to promote safety at sea. The RYA represents dinghy sailing, yacht cruising, motorboating, sportsboats, windsurfing, inland boating, powerboat racing and personal watercraft – in other words pretty much all UK watersports – and its Yachtmaster certificate is recognised worldwide as the marine qualification for captains of recreational and private vessels.

Sail training comes in many forms, but only the RYA qualifications are officially recognised

RYA Training Courses

Dinghy courses

Whilst many adults learn to sail aboard yachts, the majority of children do so in dinghies. As you would expect, the RYA offers an organised progression through the skill levels, but surprisingly it isn't all about racing. There is no doubt that a dinghy is the best place to learn about sail trimming, so whatever the age of the novice sailor, time spent in dinghies will not be wasted.

The following dinghy courses are offered by the RYA:

Level 1 – This is an introduction for complete novices that teaches you to sail in all directions, under supervision and in light winds. It also touches on launching and recovery.

Level 2 – This course assumes you have mastered the basic sailing skills from Level 1. Here you learn about rigging, launching and sailing in all directions, as well as capsize recovery and essential safety knowledge. After this course you should be able to sail and make decisions, on your own, in good conditions.

Starting them young means they should grow up to be fine sailors

Each of these courses takes two days, usually a weekend. In some schools a combined Level 1 and 2 is offered over a week, which is a good option.

After completing Level 2 you have a number of options, depending on the type of sailing you want to do. The RYA's Start Racing course is designed to give you the skills required to enjoy club racing, including how to maximise boat speed and outwit your opponents. This course will equip you to take part in club racing in good conditions.

If you aren't interested in racing, probably the next course for you is Seamanship Skills. This course covers launching and recovering the boat in different circumstances, stopping, reducing sail, recovering a man overboard and anchoring. By the end of the course you ought to be able to handle a wide range of situations afloat.

Yachting courses

The RYA's training courses cover the entire range of skills, from complete novice to professional instructor. You don't necessarily have to start at the bottom – many yachties who have sailed for years without qualifications have gone straight for the RYA Yachtmaster qualification after ensuring they are fully up to speed with all the required theory.

Complete beginners have the choice between a Start Yachting or Competent Crew course. If you don't really know if you're going to like sailing or not and just want to get a taste of what it's all about, then go on the two-day Start Yachting course. You have less to lose if sailing doesn't turn out to be your thing, but if you think it might be the course for you, days can be credited towards a Competent Crew certificate.

For those with some dinghy experience or who have sailed a little with others on larger boats, I would recommend you go straight for the Competent Crew course. You can do it over five consecutive days, two weekends if one is a bank holiday, or three weekends at other times.

In order of precedence here is a brief description of the RYA's primary courses and qualifications for sailing craft.

Start Yachting

This syllabus provides a short introduction to sail cruising for novices. By the end of the course students will have experienced helming, sail handling, rope work etc., and will be made aware of the general principles of safety at sea. The course covers:

- A basic knowledge of sea terms, parts of a boat, her rigging and sails.
- The ability to tie essential knots such as a bowline, clove-hitch, round turn and two half hitches and the figure of eight stopper knot.
- Sailing a yacht on all **points of sail**, helming and basic sail trimming.
- Learning the rules of the road and keeping a proper look out at sea.

- Learning how and where to obtain a reliable marine weather forecast.
- The dangers of, and how to recover, a man overboard.
- Your personal safety equipment – how to stay safe on board including the wearing of safety harnesses, lifejackets etc.
- An explanation of emergency equipment and routine precautions, and knowing what action to take in the event of an emergency.

Competent Crew

This basic course comprises a few days out with an instructor learning the art of daytime helming, crewing and seamanship in inshore waters. There is no written course work or examination and no previous experience is required. Subjects covered include:

- Gaining sufficient knowledge to understand orders given concerning the sailing and day-to-day running of the boat.
- A broad understanding of sea terms and parts of a boat, rigging and sails.
- The ability to bend on, reef and handle a yacht's sails.
- Rope handling, including coiling, stowing, securing to cleats and bollards, warp handling and knot tying.
- Understanding the need for and compliance with personal safety equipment including recommendations for the wearing of buoyancy aids, lifejackets and safety harnesses.
- Handling and deployment of emergency equipment including learning to operate distress flares and how to launch and board a liferaft.
- Understanding the action to be taken to recover a man overboard.
- Awareness of the hazards of fire on board a boat including precautions necessary to prevent a fire and the action to be taken in the event of fire breaking out.

Any one of the RYA's certificates of competence will stand you in good stead

- Collision avoidance regulations and all the means of keeping a proper lookout at sea.
- Learning helmsmanship, how to steer a course and the basic principles of sail trimming on all points of sail.
- Handling the yacht under power.
- A basic lesson in meteorology, knowledge of the Beaufort scale and an awareness of commonly available weather forecasting services.
- Understanding the manners and customs involved with boating, including the correct deployment of burgees and ensigns, courtesies to other craft and an awareness of the marine environment.
- Learning how to handle a dinghy.
- Carrying out general duties satisfactorily on deck and below decks in connection with the daily routine of the vessel.

Day Skipper

Part 1 (Shore based) – This is a classroom course for students with little or no navigational experience. The syllabus includes buoyage, collision prevention, chart recognition, tidal calculation, position fixing, **bearing** taking and course laying. There is no official examination at the end, but the instructor will make an assessment of your acquired skill level by giving you a test paper and issuing you with a course completion certificate.

Part 2 (Practical) – This is intended for students who are competent in the basics and have spent five days or more at sea during daylight hours. The course puts into practice the navigation, pilotage and chart-reading skills learned in Part 1, by the end of which you should be a competent daytime skipper in tidal waters. It also qualifies you for an International Certificate of Competence (ICC), which may be necessary if you want to sail or charter a yacht overseas.

Yachtmaster Coastal

Part 1 (Shore based) – This is also the theory course for the Yachtmaster qualifications and is another classroom course, this time covering more advanced navigation, meteorology, dead reckoning (DR) and estimated positions (EP), running fixes, secondary ports, lights and shapes, sound signals, passage planning and safety procedures. The course comprises around 40 hours of teaching and has three separate exam papers.

Part 2 (Practical) – Many skip this stage and go straight onto the Yachtmaster Offshore course, although those with less sea time might stick with the five-day Coastal course, which is more intensive than the Day Skipper practical. To qualify for this course entrants must have had at least 30 days sailing and 800 sea miles logged – of which 12 hours must be at night. Candidates are expected to have skippered a yacht for at least two full days and are usually yachtsmen aiming to captain a yacht on long coastal passages by day and night. The emphasis on this course is effective passage planning, day and night pilotage, yacht handling in confined spaces under power and sail, together with a thorough knowledge of handling emergency situations.

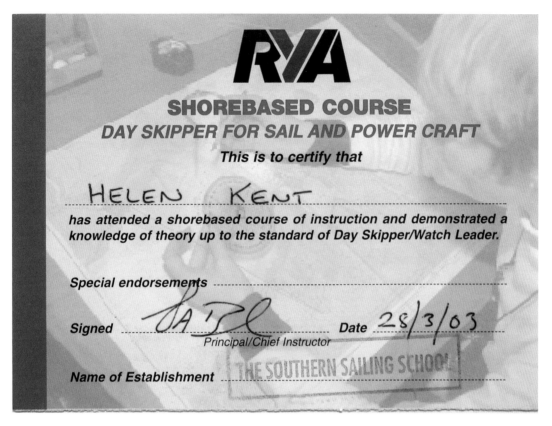

The Day Skipper shore-based course is essential if you want to sail in open waters

Yachtmaster Offshore

The Yachtmaster (YM) Offshore examination involves a fairly stringent assessment by an independent examiner after completing a thorough practical course (usually the Yachtmaster Coastal shore-based) aimed at enabling the candidates to skipper a yacht, day or night, on any passage up to 150 miles offshore that does not require astronavigation. Before taking this examination the candidate must have logged 2 500 sea miles, with five separate passages of over 60 miles, including two as skipper, two overnight and a total of 50 days at sea – half of which must have been in tidal waters.

The practical examination will include blind navigation, night pilotage, man-overboard drill, advanced boat handling under power and sail, meteorology, tidal calculations and heavy-weather tactics.

The YM Offshore teaches you all the practical essentials to skippering a yacht

To be eligible for the Yachtmaster Offshore certificate candidates must hold a valid Short-range Radio Certificate (SRC) qualifying them to operate Digital Selective Calling (DSC) VHF marine radios and be fully familiar with GMDSS emergency procedures. They must also have completed a two-day first aid course in an approved training establishment.

A commercial endorsement permits a Yachtmaster Offshore-qualified skipper to captain yachts up to 200 GRT with a maximum of 12 paying guests on board. This can be added to the YM Offshore or YM Ocean qualifications after the completion of a medical examination, a more comprehensive first aid course and a sea survival course.

Yachtmaster Ocean

This is a shore-based course with an oral examination that delves heavily into ocean passage planning, advanced worldwide meteorology and astronavigation, and it is usually taken sometime after the candidate has completed the Yachtmaster Offshore course and logged a

At the end of it all you'll feel safe and secure when you know what you're doing

few thousand miles as skipper in a variety of situations. You will be expected to have completed at least one non-stop, 600-mile passage, logged complete with sun/moon/star sights and reductions.

GLOSSARY

Aback   When the windward headsail sheet is kept tight during a tack so that she ends up hove-to

Abaft   Behind

Aft/after   The rear end or stern of the boat

Apparent wind speed   The actual wind speed as it is affected by the speed of the boat

Apparent wind angle   The direction the wind appears to be coming from when the boat is moving

Aspect ratio   A comparison of height against width, hence a high aspect ratio jib is one that is tall and narrow

Athwartships   Across the boat positioned transversely

Autopilot   A device for steering automatically

Backstay   An adjustable wire that pulls the mast aft (and bends it)

Batten   Rigid piece of battening put in the sail to keep it taut

Beam   Width of the boat

Beam reach   Sailing with the wind directly on the beam

Bear away   Turn the boat away from the wind

Bearing   Orientation in degrees from north

Beating   Close-hauled, zigzagging towards the wind

Bermudan rig   Single-masted rig with triangular mainsail

Block   Pulley for changing the direction of a line

Bolt rope   Small diameter rope sewn into the luff of the sail to hold it in the mast or foil track

Boom   Metal tube lying along the bottom of the mainsail

Bow   The forward part of the yacht

Bow-roller   Roller over which the anchor and chain are raised and lowered

Bowsprit   Spar that extends forward of the bow to support the tack of an asymmetric sail such as a cruising chute, spinnaker etc.

Broach   When a yacht gets beam-on to a wave or is over-powered by the wind and heels dramatically

Broad reach   Sailing with the wind on the yacht's quarter

Bulkhead   Transverse partition across the inside of the boat

Buoyancy   The ability to float

Call sign   Unique identification code used on VHF radio

Centreboard   A retractable keel board to reduce leeway

Centre of effort   The point at which the force of the wind acts against the sail

Chain plate   Fitting to which a shroud is attached to the hull or deck

Chart plotter   Electronic instrument showing the boat's position and course on a digital chart

Chine   Crease in the hull of a boat, usually in steel or plywood (see also: Hard chine)

Cleat   A projecting fitting for attaching a rope temporarily

Clew   The lower, aft corner of a sail

Clew outhaul   A line that flattens the foot of a sail as it is tightened

Clinker   Type of wooden boat design where the hull planks overlap

Close-hauled   Sailing as close to the wind as possible

Close reach   Sailing 60 degrees off the wind
Coachroof   The roof of the saloon
Coaming   An upstand around the cockpit to keep the water out
Cockpit   Area from which the crew control the boat
Cold moulded   Bending plywood around a frame
Companionway   The steps down into the saloon
CSM   Chopped Strand Matt: a type of glassfibre matting with fibres that are randomly positioned
Dan buoy   A small floating buoy with a flag thrown overboard to mark the position of a Man Overboard
Dead run   Sailing dead downwind
Displacement   The weight of water a hull displaces
Downhaul   A rope that pulls in a downward direction
DSC   Digital Selective Calling: the ability to direct dial another marine VHF radio, like a telephone
Flake out   Lay out rope or chain in elongated S shapes to prevent tangling
Foot   The bottom edge of a sail (see also: Loose-footed)
Forestay   A fixed wire that pulls the mast forward
Foretriangle   The triangular area between the forestay and the mast
Furler   A device that rolls the sails away
Gaff   A spar at the top of a gaff mainsail
Galley   The boat's kitchen
Galvanic corrosion   Corrosion caused by two different metals being in contact in seawater
Gelcoat   The shiny outer coating on a hull
Gennaker   A downwind sail, which is a cross between a large genoa and a spinnaker
Genoa   A headsail that overlaps the mast
Go about   Tack through the wind
Gooseneck   Hinge that attaches the boom to the mast
Goose-winged   Running downwind with the genoa on the opposite side to the mainsail
GPS   Satellite-based Global Positioning System
GRP   Glass Reinforced Plastic, or glassfibre
GRT   Gross Registered Tonnage
Grabrail   A rail on deck or down below to grab hold of
Guardrails   Horizontal wires to keep the crew on board
Gudgeon   Female part of rudder hinge
Gunwale   Pronounced 'gunnel', an upstand at the point where the deck meets the hull
Gybe   A turn where the stern of the boat crosses the wind
Halyard   A rope or wire that hoists a sail
Hard chine   A chine is a noticeable crease in the hull of a boat caused by the hull material changing direction
Harden/head up   Point the yacht closer towards the wind
Head   The top corner of a sail
Heads   The toilet or toilet compartment
Head-to-wind   The boat aligned with the bows pointing into the wind and the sails flapping
Heave-to   Stop with the jib aback and balancing the mainsail
Helm   Tiller, wheel or rudder (i.e. the boat's steering); or name for the person who is operating the steering
Hounds   The point at which the shrouds connect to the mast
In irons   Stalled in the no-go zone, the point where the yacht's sails are just flapping aimlessly

Jackstay   Wires or lengths of webbing along each side deck for the crew to clip their harnesses onto
Jammer   Deck hardware used to hold a rope under load
Jib   Headsail of no more area than 100% of the foretriangle
Jib sheet   The lines that control the tension in the jib
Jumper stay   Spar, like a spreader, on the forward edge of the mast
Kedge   A small secondary anchor
Kicking strap   Pulley system to pull down on the boom; also called the vang
Knot   Speed in nautical miles per hour
Lateral resistance   The resistance of the hull to be forced sideways in the water
Latitude   A measurement (in degrees) north or south of the equator shown on the vertical scale at the edges of a chart
Lee   The opposite direction to the wind
Lee bow   Sailing so that the tide pushes the hull towards the wind
Leech   The back edge of a sail
Lee cloth   A canvas sheet hung to retain sleepers in their bunks
Lee helm   The tendency of the yacht to bear away from the wind
Leeward   The side away from the wind
Leeway   The amount a boat slips sideways to leeward due to the wind
Liferaft   An inflatable boat with a canopy, launched when a yacht is in danger of sinking
Line   Rope that controls the sails such as a sheet
LOA   Length overall
LOD   Length over deck
Longitude   A measurement (in degrees) east or west of the Greenwich meridian shown on the horizontal scale at the top and bottom of a chart
Loose-footed   A sail that is attached by the tack and clew only
Luff   The leading edge of a sail
Luff up   To turn towards the wind so that the sail just starts to flap gently
LWL   Waterline length
Mainsail   The sail set on the mainmast and boom
Mainsheet   The line that controls how far the boom is pulled in
Mayday   Raising the alarm when there is grave and imminent danger to a vessel or person
MOB   Man Overboard
Nautical mile   1 nautical mile (nm) equals 1.15 statute miles
No-go zone   The area 40 degrees either side of the true wind direction
Osmosis   Waterlogged glassfibre lay-up causing blisters to appear
Outhaul   A rope or wire that pulls the foot of the mainsail towards the aft end of the boom
Passerelle   A gangplank used for getting off the stern of a yacht onto the shore
Pitch-pole   When a boat capsizes end-over-end
Pintle   Male part of a rudder hinge (see also: gudgeon)
Point of sail   Close-hauled, reaching or running
Poled out   Holding the clew of a sail out with a spinnaker or whisper pole
Port   Left-hand side of the boat
Port tack   Sailing with the wind on the port side of the boat
Preventer   A rope system for holding the boom out when the yacht is running and preventing a gybe should the wind back the sail

Pulpit   The stainless steel structure at the bow
Purchase   Mechanical advantage of a block and tackle
Pushpit   The stainless steel structure at the stern
Quarter   The aft corners of the yacht
Rake   The angle that the mast leans from the vertical
Range   Distance from the yacht to her destination
Reaching   Sailing between 60–120 degrees off the wind
Reef cringle   A metal eye through which the reefing ties run
Reefing   Reducing sail area
Rhumb line   A straight line to the destination
RIB   Rigid Inflatable Boat
Rig   General term for the mast, spars and sails
Rigging screw   Bottle screw adjuster for tensioning the standing rigging
Roach   The top part of the mainsail aft of a direct line between the head and clew
Rocker   Fore and aft curve of the hull
Running   Sailing before the wind (i.e. away from it)
Running backstay   Adjustable lines from the mast (usually at the head of the inner forestay) to the yacht's quarters; often just called 'runners'
Running by the lee   Sailing downwind with the wind immediately behind the boom (i.e. about to gybe)
Running rigging   The part of the rigging that controls the sails including the halyards, sheets and guys etc.
Sail-tie   A piece of webbing tied around a sail after it has been lowered
Sail trim   Setting the sails so that they use the wind most efficiently
Saloon   The main cabin area inside a yacht
Seacock   A valve in a pipe through the hull, to let water in or out
Shackle   A metal hoop and pin used to attach one object to another
Shroud   A wire that supports the mast sideways
Skin fitting   A water pipe fitting through the hull
Slab reef   A method of reefing the mainsail whereby a whole portion of the sail is taken down in one go
Slot   The gap between the jib and the mainsail
Snuffer   A large sock that is pulled down over a spinnaker or cruising chute
Sole   Floor
Spar   General term for a mast, boom or pole
Spinnaker   A large downwind sail
Spreader   Horizontal arm projecting from the mast to give a shroud more leverage on the mast
Spring   A mooring rope that stops the boat moving forwards or backwards
Stanchion   A vertical rod that supports the lifelines
Standing rigging   The part of the rig that supports the mast (i.e. the shrouds and stays)
Starboard   The right-hand side of the boat
Stem   The forward extremity of the boat
Stern   The aft extremity of the boat
Tabernacle   Mast-foot support for a deck-stepped mast
Tack [1]   A turn where the bow of the boat turns through the wind
Tack [2]   The lower, forward corner of a sail
Telltale   A streaming piece of cloth indicating the airflow over a sail
Tiller   Pole connected to the rudder for steering a boat

Toe rail   A raised rail around the edge of the boat where the hull meets the deck

Topping lift   A line used to lift or lower the boom

Transom   The transverse after end of a boat

Traveller   A slider on a track used to adjust the effective pivot point of a sheet

True wind   Wind relative to a stationary object

Twist   The difference in the angle of the sail to the wind at the top, when compared to the angle at the bottom

Unstayed   A self-supporting mast with no standing rigging

Uphaul   A line that pulls objects upwards such as the spinnaker pole

Upwind   Any course closer to the wind than a beam reach

Vang   A pulley system to pull down on the boom; also called a kicking strap

VMG   Velocity Made Good (i.e. the actual speed towards your destination or waypoint)

Wake   Trail in the water left behind a moving boat

Waypoint   A mark on an electronic chart indicating the boat's final or intermediate destination

Weather helm   Condition when sailing to windward when the helm has to be held up to the direction of the wind, usually caused by an imbalance of sail areas or poor trim

Wetted surface   Underwater area of hull

Whisper pole   A lightweight pole used to pole out the jib when goose-winging

Winch   A mechanical device to help wind in a rope, using a winch handle

Windex   A pivoted arrow at the top of the mast indicating the wind direction

Wind gradient   The difference in wind speed between sea level and masthead

Windlass   A winch for winding in the anchor chain

Windward   The direction from which the wind is blowing